THE CONCEPT OF
THE POSITRON

THE CONCEPT OF
THE POSITRON

A PHILOSOPHICAL ANALYSIS

BY

NORWOOD RUSSELL HANSON

B.Sc., M.A. (Cantab), D.Phil. (Oxon)

Professor of History and Logic of Science
Indiana University

CAMBRIDGE
AT THE UNIVERSITY PRESS
1963

CAMBRIDGE UNIVERSITY PRESS
Cambridge, New York, Melbourne, Madrid, Cape Town, Singapore,
São Paulo, Delhi, Dubai, Tokyo, Mexico City

Cambridge University Press
The Edinburgh Building, Cambridge CB2 8RU, UK

Published in the United States of America by Cambridge University Press, New York

www.cambridge.org
Information on this title: www.cambridge.org/9780521106467

First published 1963
First paperback printing 2010

A catalogue record for this publication is available from the British Library

ISBN 978-0-521-05198-9 Hardback
ISBN 978-0-521-10646-7 Paperback

To

LESLIE *and* TREVOR

CONTENTS

The plates face pp. 137, 178 *and* 217

ACKNOWLEDGEMENTS

The following work results from so much assistance by individuals and institutions that fully to tell how much would require another 225 pages. The most I can hope for here is accuracy and completeness in my recognition of this debt. The author's gratitude, which laces together all the following entries, must be understood by the reader.

To Professors Anderson, Blackett and Dirac—the principals in the story—I extend my warm thanks. My only hope is that inaccuracies in this book will not outweigh their own splendid efforts to convey to me the excitement of the birth of the Positron concept.

Further assistance was liberally provided by Professors J. R. Oppenheimer, N. Mott, H. Bethe, D. Skobeltzyn and D. Wilkinson. Invaluable suggestions were provided by Professors Bohr and Heisenberg, as well as Lamb, Rosenfeld, Joliot-Curie, Jeffreys, Konopinski, Hill and Yennie—with 'assists' by Drs Newton, Langer, Woodruff and Taylor. Amongst my philosophical friends, Professors Ryle, Feigl, Maxwell, Sellars, Toulmin, Grünbaum, Feyerabend, Körner, Hempel and Putnam did their very best to help the style and arguments of this book to approximate to readable and reasonable prose.

Dr John Ziman read the manuscript on its way to the Press and worked with Mr Becher in a manner which improved the structure of the book.

Even at the promissory stage of this research—back in 1953 and 1954—several Foundations helped in their inimitable way. The Nuffield, Ford, and Rockefeller Foundations lubricated the research machinery at some rather sticky moments. The Minnesota Centre for Philosophy of Science provided discussion and facilities which were most helpful. The Indiana University Foundation, and the Research Council, provided for typing and secretarial assistance which accelerated the project. My thanks also to St John's College, Cambridge, and to the Master, for their many forms of assistance during the past ten years. I am indebted also to the United States Air Force and to the American Philosophical

Society for travel grants made available at critical moments. Amongst my many friends and colleagues I should like especially to mention Professors A. R. Hall and M. B. Hall, R. Buck, M. Scriven and E. Grant for the moral sustenance they so often provided. Paul McEvoy helped with the notes and index. And to my dear wife and children, Fay, Leslie and Trevor, I extend my loving gratitude for their infinite patience.

NORWOOD RUSSELL HANSON

Reference numbers refer to the notes on pp. 184–225.

INTRODUCTION

The denouement of this book is its ninth chapter. There the intricate story of the discovery of anti-matter is set out. Each chapter preceding this attempts to secure some philosophical point, without which features of the positron discovery would be difficult to grasp.

In chapter I, 'Light', the conceptual basis of Newton's Theory of Fits is examined: did the 'crucial' experiments of Young, Fresnel, Fizeau and Foucault demolish Sir Isaac's modified corpuscular theory? Here two steps are taken towards the idea of the positron: first, the historical foundation and conceptual superstructure of the wave-particle duality are set out; secondly, the logical structure of the *experimentum crucis* is exposed. Without these, the theoretical basis and the experimental support for the positron hypothesis could hardly be appreciated. [See appendix I.]

Chapter II re-explores the dichotomy between 'Explaining and Predicting'; it is suggested that Hempel's thesis concerning the logical symmetry between these two concepts is inadequate to any description of microtheory. This conclusion constitutes a further step toward understanding the positron. The 'hole theory' of the positive electron is an explanation of things like pair creation and annihilation; none the less, as a matter of principle, this theory cannot predict when any given pair will be created. The relationship between Dirac's algebra and the Oppenheimer-Blackett expositions of the 'hole theory' must be traced with care to determine where Hempel's thesis is valuable yet also misleading.

Chapter III concerns 'Picturing'. Here we explore why classical dynamical and geometrical models of fundamental particles are untenable in principle. This discussion attempts to place the 'hole theory' into perspective, and suggests why the physical properties of the positron can never be formed into a picture, or into a nineteenth-century type of model. [See appendix II.]

Chapter IV forces a logical confrontation: 'The Correspondence Principle and the Uncertainty Principle'. There is an acute conceptual tension within quantum theory between these principles, and the consequences of this affect our understanding of the positron itself.

Chapter V, on 'Interpreting', purports to defend the 'Copenhagen' interpretation of quantum theory. This issue has been most exciting in recent philosophy of science. The view expounded here has not gone unchallenged: it puts forward reasons why the theses of Vigier, Bohm and Feyerabend should not perhaps be weighted as heavily as their authors wish. In any case, some sympathetic understanding of what motivated the Copenhagen interpretation is an essential part of the conceptual background to the discovery of the positive electron.

Chapter VI elaborates points within the preceding chapter. Under the title 'Some Cautions' we raise further difficulties about the Vigier–Bohm–Feyerabend approach, and urge restraint lest historical misunderstandings arise and engender conceptual ones. The one with which we are most concerned, of course, involves the positron itself.

The next chapter (VII) discusses 'Uncertainty' again. We examine some standard 'counter-instances' to the Uncertainty Relations: each of these collapses on analysis; the Uncertainty Relations are delineated as *the* conceptual foundation stone of quantum theory. Dirac's 1928 paper on the relativistically invariant, spinning electron cannot even be comprehended if the Uncertainty Relations are construed as anything less than a *theoretical* boundary condition of quantum physics. Attempts to treat the relation $PM - MP = (h/2\pi i)\psi$ *merely* as an observational limitation are shown to fail.

Chapter VIII is called 'Equivalence?'. The 'proofs' of Eckart and Schrödinger, to the effect that the Wave Mechanics and Matrix Mechanics of 1926 are equivalent, are, it is suggested, faulty in their conclusions. This chapter has stimulated some comment among theoretical physicists, and its thesis appears to be substantiable. This is in itself quite important; the 1928 paper of Dirac constitutes not simply algorithmic ingenuity, but a redesign of the very conceptual framework of then-extant quantum theory. The Dirac formalism achieves what the Schrödinger and Eckart 'proofs' could not achieve. Wave Mechanical and Matrix Mechanical solutions are both easily generable within the Dirac notation; hence this constitutes a stronger claim for the theoretical equivalence of Wave Mechanics and Matrix Mechanics than when they are represented in the original Schrödinger and Heisenberg notations. Since the positron concept derives directly from the innovations of the 1928 paper, our rejection of the proven equivalence of Wave Mechanics and Matrix Mechanics before that date constitutes an essential last step toward the discovery of the positron idea. [See appendix III.]

Chapter IX is called 'The Positron'. Here some of the intertwining strands of theory, observation, and experiment in microphysics are disentangled, with the result that we can distinguish three different discoveries within this single complex. I arbitrarily designate these 'the Anderson particle', 'the Dirac particle', and 'the Blackett particle'. The first was a truly remarkable observational discovery based on strict reasoning, but largely innocent of any complex theoretical considerations. The second constituted a dramatic theoretical advance—but one carried on, at first, in the abstract realm of the *gedankenexperiment*. Although wholly independent of each other, these first two discoveries both rebounded from a reluctance to entertain any particle except the negatron and the proton. One function of this chapter is to explore such reluctance; its roots go back into the history of electrical theory, and especially into late nineteenth-century electrodynamics. The third discovery, Blackett's, is characterized as a '*metaphysical*' discovery. In addition to constituting a primary advance in itself, Blackett's work is important for recognizing that the prior discoveries of Anderson and Dirac were the *same* discovery. From quite different directions Anderson and Dirac had *backed* into similar conclusions about the same material entity. But that this was so was not known until Blackett published his paper of 1933. [See appendix IV.]

The author's own concern with the positron concept began in 1946. As more and more has been learned of the total story, deeper analysis has sliced into each new speculation and attitude. Special conjectures of my own have been tested and modified, and usually abandoned, in conversations and correspondence with theoretical and experimental physicists, with historians of science, and with logicians. To them all I am grateful.

The positron is the first anti-particle. It was the first genuine alternative to the dominance of particle theory by the negative electron and the proton; it is at once a reluctant cause and the dramatic result of the twentieth-century conjecture that matter can be created out of energy. Many related problems of interest to historians and logicians of science literally spill from this cornucopia of physical concepts.

CHAPTER I

LIGHT

A

During the last half-century startling discoveries have been made by physicists: these have affected our understanding of heat, light, and electricity, in ways yet fully to be realized. None the less, the conceptual structure of the findings of Planck, Einstein, Compton and Dirac is by now beginning to emerge. Let us consider this structure against the three centuries of physics which preceded it, with particular reference to Newton's optical theory and the famous 'crucial experiments' to which it led.

The disturbing discoveries of the present century have concerned the notion of radiation as a continuous transfer of energy. This is just the notion and the controversy which had earlier centred on the nature of light. Was the movement of light from sun to earth an unbroken propagation of energy, analogous to rolling surf-waves? Or did it consist in a discontinuous emission of discrete packets of energy, like the peppering of a target with bullets?[1] The work of Planck, Einstein and Compton, all of it theoretically persuasive and experimentally ingenious, supported this latter conclusion.

The orthodox nineteenth-century position concerning radiant energy was elegantly and forcefully expressed in the electromagnetic wave theory of James Clerk Maxwell and H. A. Lorentz. But this position was assailed by difficulties even before Planck. Certain calculations[2] required infinite values for the total density of energy; an inconceivable state of affairs. Things worsened when Max Planck studied how hot black bodies give up and take on radiant energy. He conceived of a beautifully designed furnace, and sensitive detectors. With these he established that any such body emitted energy not continuously, but in distinct, discrete pulses: there were calm valleys between emitted pulses; intervals during which little or no radiation left the hot, black body. Planck compromised with the orthodox theory: he conceded that bodies *took in* radiation in a continuous way; but his experiments forced

5

him to assert that they gave it off in equal pulses. Energy absorption was continuous; energy emission discontinuous. Planck preserved the continuity of radiation, for this alone seemed compatible with the requirements of the highly confirmed wave theory; he restricted his heresy to emission of radiation from hot bodies. Hence, Planck's theory of 1901.[1]

In 1887 the photo-electric effect was discovered by Hertz. If we charge an electroscope negatively, so that the gold leaves mutually repel each other, and then cover the upper surface with an alkaline metal (e.g. zinc), a remarkable phenomenon occurs when the apparatus is bathed in X-radiation. The electroscope loses its charge and the leaves slowly fall together. Nothing in classical electromagnetic theory accounts for this: X-rays lack charge; why should the electroscope lose charge when X-radiated? In 1905 Einstein boldly supposed X-radiation itself to be composed of the discrete pulses of energy of which Planck had spoken:[2] the action of these pulses on the matter constituting the electroscope might explain the effect. Einstein developed the idea of the photo-electron as an ordinary matter-electron, photo-electrically expelled from an electroscope. Whenever an energy pulse crashes into an electron within the matter composing the electroscope, the result is interpretable as a classical two-body problem analogous to billiard balls under impact.[3] Occasionally, however, one of the matter-electrons will be knocked out of the electroscope, carrying its charge with it; as this process continues, the total negative charge of the electroscope slowly disappears.

A. H. Compton made a further discovery in 1923 which established that, whatever else it may also be, light radiation is indisputably particulate.[4] By playing X-rays on to a carbon block and then trapping the reflected radiation in a sensitive detector, Compton revealed that the scattered rays were not all of the same frequency as the original beam. Some reflected rays had lost energy in the transaction. Again, Compton explained this by supposing a classical two-body interaction between a photon and one of the electrons in the carbon. This effect, however, differs strikingly from the photo-electric effect; here the light-corpuscles are not annihilated by their contact with matter: they bounce back, with a classical interchange of energy as between billiard balls.

Each of these discoveries bred controversy concerning the

'proper' interpretation. That radiation should be particulate and discontinuous was too startling 40 years ago for most physicists to countenance.

Indeed, this double-stranded line of development becomes beautifully knotted in contemporary experiment. If, in an electron microscope, a beam of electrons emitted by a filament is interrupted by a plate with a tiny hole in it, a vague splash of light will appear on the target screen beyond. This is caused by the electrons coming through the hole and then spreading out. Now make another hole very close to the first. Do we see two patches of light? No. We see the same patch, only now much brighter, and striated with several parallel dark bands. Choose now one of the brightest parts of the patch on the screen: which of the two holes do the electrons reaching that spot pass through? Cover up one of the holes and see what happens. The bright-dark striations vanish at once, and the intensity of light at the selected point falls off sharply. But this does not mean that most of the electrons were going through the hole we covered, since precisely the same effect would have been detected had we covered the other hole. The striations and the extreme brightness, in other words, are observed only when *both* holes are open. In these circumstances we simply cannot observe through which hole any particular electron passed. So here the result of the experiment is affected by any such attempt to follow it in finer detail than it seems prepared to allow. From the instant the electrons leave the filament until they impinge on the screen we are denied the luxury of looking at them: to observe is to transform.

Yet how to describe what *is* observed? The electron particles are clearly interfering in some 'classical' wave-like manner. And the resultant illumination is obviously also particulate in some fairly fundamental way. The same startling phenomena mentioned earlier seem to intertwine in this electron-microscope example.

B

Why should these findings have been startling? Nature is what it is; why should men be alarmed to learn that light is as much like a hail of fine sand as like the jiggling of a clear jelly?

The answer is familiar. Men are surprised by nature because

they themselves make it impossible not to be surprised. Scientists make decisions concerning what certain phenomena must be like, and are then startled to learn that nature refuses to co-operate with their ideas. Sometimes we behave inflexibly in these contexts; more than one scientist has had difficulty in conceiving that nature could be other than he originally supposed it to be. To overcome this has been the province of genius.

From earliest times it had been known that certain properties of light are best explained by supposing it to be corpuscular. For example, geometrical optics, that science which examines relationships between shadow-lengths, the positions of light-sources, and the heights of illuminated objects, was established in Greek antiquity. Thales is said thus to have determined the height of the pyramids. But before such a science can be formulated, shadows must be assumed to be sharp, i.e. not to blur off indeterminately. Had this not been so, Thales could not have known from where to measure the pyramid's shadow. But if one considers the 'shadows' which water waves cast behind a pier, they are not at all sharp; instead of seeing a calm zone separated from the waves by a sharp boundary, the latter curl round behind the pier. Sometimes they obliterate the 'dead zone' entirely. So waves did not seem, to the ancients, a good model for light propagation. Since the shadows cast by the sun were sharp, it seemed that sunlight travelled along perfectly straight lines. Euclid, indeed, makes this the basis of his optics,[1] and of all his studies of perspective.[2] This is consistent with the idea that light resembles more a spray of fine particles than it does undulations within a thin, clear jelly. Even so, Euclid himself opposed the 'emission theory' of Pythagoras. Sand blown obliquely against a book leaves a sharply demarcated streak on the far side. Empedocles argued along these lines for a particulate theory.[3] This Principle of the Rectilinear Propagation of Light is immediately established when one considers mirror reflexions which are easily explained on the rectilinear principle and the particulate theory of light. Claudius Ptolemy[4] traces the rectilinear motion of a reflected ray by faithfully comparing it with an object thrown against a wall; and he continually speaks of a 'slinging action'. All this itself became the subject-matter of a distinct science—catoptrics—in connexion with which the great names of Euclid, Hero, Ptolemy and Alhazen stand forth.

LIGHT

The particulate theory was virtually the only one concerning the nature of light until the early seventeenth century.[1] Then Grimaldi studied the shadow cast by a hair.[2] This shadow was fuzzy at its edges: indeed, as the illumination intensified the fuzziness broke up into a series of fringes or stripes running parallel to the shadow's edge. Grimaldi made the important observation that these fringes appeared not only outside the shadow's edge, but actually fell within the shadow itself. This is crucial: we shall refer to it again.

Even before Grimaldi's observation, wave motion had been studied. It was suspected that periodicity in a phenomenon, any kind of a regularly recurrent pulse, or waxing-and-waning, indicated that the event was wave-like in nature. Thus the beats emitted in the lower registers of a cathedral organ distinguished sound as consisting in a wave-like propagation of energy through air.[3] The high spots and dead calms which evenly intersperse where, on the surface of a pond, two wave fronts intersect and overlap, were construed as the same kind of phenomenon. Even things like magnetic effluvia were suspected of having this basic property. Thus Grimaldi's observation carried the suggestion that perhaps light too had a wave nature.[4] Further work by Descartes, Huygens, Newton, Euler and Hartley, brought extensive observational and theoretical support to this thesis, to which we must later return.

For the moment, it is relevant to consider the researches of Young and Fresnel which, in the early 1800's, established light as a periodic disturbance of some sort. They showed light to interfere, constructively and destructively, just as do water waves and sound waves. Young with his renowned two-slit experiment,[5] and Fresnel, with his equally famous bi-prism experiment,[6] were able to bring two distinct, yet physically identical, wave fronts of light into overlapping contact. These were made to run across each other at a small angle, just as two wave fronts of water, or air, can be made to cross and interfere. The remarkable result was that both Young and Fresnel were able to reveal on their target-screens bands of bright light alternating with dark patches.[7] From this, given that detecting such periodicity indicates the presence of wave phenomena, the Young–Fresnel experiments proved that light must consist in an undulatory propagation of energy. Nor has anything since discovered disproved their findings. But, for historical

9

reasons to be examined, the results of their work became known not only as that which established the undulatory character of light: it was taken also to *disprove* the theory that light was in any way particulate.

Later experiments, by Fizeau(1849)and Foucault (1850), carried this logical progression further. The particle theory of Newton, La Place and Biot[1] requires that the velocity of light should increase as the radiation passes into a denser medium. Fizeau and Foucault established that this did not happen: by an ingenious experimental arrangement Foucault was able to disclose that the velocity of light in water is less than its velocity in air. This did seem crucial against the corpuscular theory. No consistent theory could allow the velocity of light in water to be at once greater than and less than its velocity in air.

A logical monument was erected by the wave theorists to commemorate this 'defeat' of the corpuscular theory. One must mention here the names of Poisson, Green, MacCullagh, Neumann, Kelvin, Rayleigh, Kirchhoff, and last, the great James Clerk Maxwell who developed the theory to a high order of precision and elegance, and applied it in totally unsuspected areas. In all this wave-theoretic work the very concepts of *particle* and *wave* came to be fashioned in logical opposition to each other. Particle dynamics on the one hand, and electromagnetic wave theory on the other, became fashioned as mutually exclusive and fundamentally incompatible concept-systems which, between them, could embrace every type of energy transfer. Even now we cannot easily conceive of a third way of propagating energy. Still, the two theories could never be applied simultaneously to the same phenomenon. A particle is a dynamical entity with sharp co-ordinates, it is in one place at one time; no two particles can share the same place at once; when particles collide there is a familiar impact and rebound. A wave disturbance, however, is fundamentally lacking in sharp co-ordinates; in principle it spreads boundlessly throughout the volume of the undulatory medium.[2] Contract a wave to a point and you destroy it; indeed, 'Wave motion at a geometrical point' gestures towards an inconceivable state of affairs.

Moreover, one can sensibly speak of two waves being in the same place at once; this is clear from observing two surf-waves crossing at a point, there generating either crests, calms or troughs, and continuing on as before. Nor is anything in wave motion strictly comparable with events like collision, impact, recoil, and the kinetic transfer of energy. So obvious was this to nineteenth-century thinkers, and so precise its mathematical expression in the work of Maxwell and Lorentz, that one could say that if any disturbance displayed *wave* properties P_1, P_2, P_3, then any similar, but particulate, disturbance would involve the very opposite properties; *not* P_1, *not* P_2, *not* P_3. It was unthinkable that any event should be at once describable by both classes of predicates; the suggestion itself seemed absurd. This means not merely that nobody had yet been able to picture such an event; in the only notations available for describing particle and wave dynamics such a joint description would have constituted a plain inconsistency. The wave concept and the particle concept were now at opposite notational poles.

C

At this stage in the history of science, not before, the earlier experiments of Foucault, Fresnel and Young, came to be spoken of *ex post facto* as crucial. They were crucial experiments because, for late-nineteenth-century thinkers, they sharply decided the issue between the wave theory and the particle theory.

Before pursuing this let us recall our earlier question: why were the discoveries of Planck, Einstein and Compton so startling? We are now in a position to answer: because, given the conceptual preparation just discussed, a proposed granular character for radiation would appear not only unusual, but virtually unintelligible. A physics in which the concepts of *particle* and *wave* have been designed in logical opposition can hardly rush to embrace the discovery that, in addition to properties revealed by Young, Fresnel and Foucault, light radiation must also—and at the same time—be regarded as particulate. To the orthodox nineteenth-century physicist, entertaining this idea would have been like thinking of a quadrilateral triangle, or an intangible physical object. Black body radiation, the photo-electric effect, and the Compton effect, were startling because physicists had already set their minds against the possibility of such events. The historical steps involved in their so setting their minds must now be considered in detail.

The Young–Fresnel experiments were vaunted by nineteenth-century wave theorists as crucial against Newton's corpuscular theory. Why? By devices already outlined, both Young and Fresnel ran identical beams of light across each other at shallow angles. If light were *really* wave-like, then at certain points wave crests in the two beams should coincide, and at other points the waves of one ought to coincide with the troughs of the other. The result on the target-screen should be brilliant streaks at the crest-intersections and little light, if any, where the troughs cancel out the crests. Because this phenomenon is actually observed, the physicist concludes, to the discomfort of the deductive logician, that therefore light is wave-like in character; i.e. if the antecedent obtains then the conclusions should describe observations; the conclusions do describe observations, therefore the antecedent obtains. Now in as strong a sense of 'proof' as inductive science can ever supply, these experiments prove the undulatory nature of light. But they certainly do not prove that light cannot also be particulate.

This last could be inductively established, only if, in addition to the principle: *if X exhibits periodic behaviour, then X is wave-like*, one also pronounced the further principle: *every energy transfer in non-solid media from a distant source is either effected in a wave-like or in a particulate manner, but in at least one of these, and in no case in both at once.*[1] Logicians will recognize this principle to turn on the 'exclusive' interpretation of the connective 'or'. When we say '*P* or *Q*' and mean this in an exclusive way, this means '*P* or *Q* obtains, at least one obtains, and in no case do both *P* and *Q* obtain'. This is to be distinguished from another use of disjunction, that termed 'inclusive'. A disjunction interpreted inclusively may be put thus: '*P* or *Q* obtains, at least one, and very possibly both'. This difference can affect interpretations of the meaning of empirical claims. If I refer to someone as being either a physicist or a chemist, I do not rule out the possibility that he may, like Urey or Pauling, be both a physicist and a chemist. If, however, I refer to someone as a man or a woman, I do not in general leave it open that the person might be both man and woman. If I say Jones is a Harvard or a Yale man, I need not mean that he cannot be both. But if I say Jones is in Cambridge or in New Haven, I do, generally, intend that he cannot be in both places at once.

The Young–Fresnel experiments are crucial against Newton's corpuscular theory only if set within an exclusive use of this disjunction: 'light is particulate or undulatory'. Though Young and Fresnel never made this explicit, their work can be crucial only if one presupposes the principle: 'It is not possible for any energy-propagation both to be wave-like and particulate'

$$[\sim \Diamond (\exists\ x)(W_x.P_x)].$$

Only then can one infer from 'Energy-propagation X is wave-like' to 'That same energy-propagation is not particulate'. This latter inference is the logic of the crucial experiment. Yet the conclusion follows only from both $(\exists x)(W_x)$ and $\sim \Diamond (\exists x)(W_x.P_x)$. So unless the principle is accepted the crucial experiment cannot be.

What Young and Fresnel described, they actually saw: no one, not even a resurrected Newton, could say they misdescribed the facts. But one could argue, as Newton would have done, that nothing in the Young–Fresnel experiments forces one to accept an exclusive disjunction between particle concepts and the wave concepts. Newton might have said: 'Yes, your experiments establish that light radiation consists in some kind of undulation. So it must be both wave-like and corpuscular.' Nor could Young or Fresnel insert anything into their experiments *ex post facto* which would rule out such a conclusion; not without being arbitrary in their decisions. This is so despite the fact that Maxwell's followers would have found Newton's conjecture unintelligible. But for Newton himself this possibility was far from unintelligible: we must now see why.

D

Sir Isaac Newton was no stranger to the hypothesis that light consists in undulations. Consider the phenomenon known as 'Newton's rings', the concentric, rainbow-coloured haloes observed when two thin plates of glass, of slight opposite curvature, are superimposed. Newton observed these rings carefully: he even introduced some ideas of wave theory to explain their presence. He also knew of Grimaldi's observations,[1] although their essential feature escaped him; viz. the fringes on the inside of the hair's shadow. The proof of this is in *Opticks*, III, pt. I, obs. I: 'And it's

THE CONCEPT OF THE POSITRON

manifest that all the light between these two Rays...is bent in passing by the Hair, and turned aside from the shadow..., because if any part of this light were not bent it would fall on the Paper within the Shadow...contrary to experience.'

The outside fringes Newton must have explained by his 'composite' theory of waves and particles. These corpuscles were somehow influenced by a concomitant wave disturbance which distributed the deflected pellets in an orderly, periodic manner.[1] But beyond all this, Newton firmly embraced the Principle of the Rectilinear Propagation of Light. For him this principle was conceptually incompatible with the requirements of wave motion.[2]

Newton often observed a bright light throwing a sharp shadow from the edge of a razor, in a perfectly rectilinear manner, across five feet of open space—leaving on the opposite wall a well defined silhouette of the razor. He could not conceive this as the effect of the light undulating like a water wave. 'Light waves' obviously did not spread in the way in which any intelligible wave theory of Newton's time would seem to require. *Ergo*, light waves *simpliciter* did not exist. Newton argued similarly for the case of a razor reflected in a mirror. On seventeenth-century wave-theoretic accounts of reflexion, what we see in the mirror should be diffuse. Refraction raised a similar consideration. Nor is this argument unsound: Professor Drude writes in *The Theory of Optics* (1902):

...in the case of a very small opening (in a diaphragm) the light is spread out behind (the opening) upon the screen so far *that in this case a propagation cannot possibly be rectilinear* [p. 1].

Drude is, in effect, arguing Newton's point from the opposite direction. Since light *is* wave-like, he says if we restrict the beam to fine dimensions, the result will be wave-like and hence *not* rectilinear propagation. Drude continues:

It is altogether impossible to isolate a single ray and to prove its physical existence. For the more one tries to attain this end by narrowing the beam, the less does light proceed in straight lines, and the more does the concept of light rays lose its physical significance [p. 2].

Thus both Drude, the wave theorist, and Newton the atomist, argue that what is wave-like cannot be propagated rectilinearly. Drude claims that because light is wave-like, therefore a point source of light cannot, despite appearances, give rise to perfectly rectilineal propagation. Newton argues that because shadows are sharp, and because sun rays come over a wall in straight lines, therefore light cannot be wave-like. Both arguments have the same logic, but proceed in opposite directions.

However, Newton had difficulty with the fact that from a single point on a smooth, transparent surface, some of the incident light is reflected, while some of it refracts through the medium.[1] Thus, for example, looking down into the River Cam, Newton could at once see the surface-reflexion of the sun overhead, and also view the river's bottom illuminated by that same sunlight. The reflexion consists in light 'bounced off' the surface, and the illumination of the bottom consists in light passing through the surface of the river. Any simple corpuscular theory of light is in trouble as a result of this observation. Newton knew of no other natural situation wherein particles shot at a surface are partially reflected and partially refracted, without damage to the reflector-refractor.

Newton's answer to this was bold, and adequate to the facts as he knew them, but we must make neither too much nor too little of his conjecture. The idea contained in his 'theory of fits of easy reflexion and easy transmission of light' is simple. Newton considers a pebble thrown into a clear, still, deep pond. Immediately the pebble strikes, a circus of waves on the surface—and in the depths—of the pond results. We watch the pebble, brightly illuminated, as it descends. But we view its path from above the surface, while this ripples beneath our gaze. The pebble appears to move down and then halt, and then descend and halt again, all the way to the bottom. Regard the pebble as a pure particle: its descent is still observed as if it consisted in a series of fits and starts, depending on whether or not the surface-wave through which we view the pebble is moving so as to make the stone appear to progress downward, or otherwise. [See appendix I.]

This is only crudely analogous to what Newton actually means. His idea is as follows: as a particle of light traverses the ether, the latter undulates as did the pond's surface when broken by the pebble. Newton's ether waves, however, move more quickly than the light particle itself. The particle's motion is felt immediately throughout the ether, as a kind of vibratory pressure. This 'pressure' is evident not only behind the particle, and at its sides, but also far in front—much as the ice will crumple yards ahead of the ice-breaker's bow. This effect is always manifested in a wave-like way. The picture, then, of any light-particle as it infringes on a smooth reflecting surface is this: either it arrives at the surface on the crest of an ether wave—in which case the particle's propensity to continue straight through the surface will be enhanced—or it arrives between ether waves, in which case further forward progress tends to be impeded. There are no other possibilities. These crests generate in the particle a disposition to be transmitted, i.e. to enter into the body of the water and be refracted by it. If the particle arrives between crests, however, it will have a disposition to be reflected. These dispositions are the so-called 'fits'.[1]

That is a simple exposition of the theory of fits of reflexion and transmission. The theory offered not only an explanation of how reflexion and refraction can occur at the same spot on an optical surface, but also coped with more complicated optical effects encountered in haloes resulting from the superposition of thin plates, and other phenomena besides.

Thus Newton would have rejected the hidden assumption of Young and Fresnel. The entire point of his Theory of Fits, for us, is that it reveals Newton's refusal to let preconceptions bully him into explaining phenomena in any form other than as they actually appear. If the course of nature obliges him to mesh particulate and wave concepts when discussing light, so much the better for man's respect for the complexities of the phenomena. How ill it becomes nineteenth-century wave theorists (e.g. Stokes and Mach) to write off Newton's theory as an impossible compromise between incompatible notions. He himself undertook the 'compromise' only because the data, as he knew them, revealed themselves as a *de facto* compromise. Newton's rings and rectilineal propagation: how else could the unswerving seventeenth-century empiricist describe the entity—light—which so manifested itself? How else other than in the very theory Newton invented? Newton's *Opticks* is brilliantly adequate to the phenomena as he knew them in the seventeenth century. He does not explain Stokes's and Mach's problems because he never had them.

If, however, nineteenth-century commentators seem unhistorical in their negative reflexions on Newton's *Opticks*, the same may be said of many Nobel-prize winning enthusiasts of our century who see in Newton's conjectures the prophet's vision of Planck, Einstein and Compton. This thesis is also objectionable: just like the Victorian wave-theorist's position, it fails to understand what Newton's problems were in terms of Newton's own data. It should be obvious that these data were qualitatively and quantitatively different from those which perplexed physicists two centuries later. Newton never had the problems of Bohr and De Broglie either: why should his seventeenth-century insights be hailed as intuitions of answers to our twentieth-century difficulties? Newton's greatness lies in the profound insight he had into his own problems—something ignored both by his nineteenth-century critics and his twentieth-century idolators.

So, the discovery by Young and Fresnel that light is wave-like in some respects, would not, by itself, force Newton to abandon the complementary hypothesis that light is also particulate. The experiment of Foucault, however, is different. It does not deal directly with whether or not light undulates: it is concerned only with the derivative question of the velocity of light. The experiment is crucial to the whole issue before us in a way the Young–Fresnel experiments never were.

One thing definitely entailed by Newton's corpuscular theory is that light should accelerate on entering denser media.[1] The point comports with the mechanical (atomistic) natural philosophy of the times, itself due in no small part to the success of Newton's work in mechanics. Medieval physics had been too full of ineffables; not only did Scholastic 'forms' and 'essences' befog every scientific inquiry, but slippery notions of influences, propensities, and effluvia, regularly confounded fourteenth- and fifteenth-century discussions of light and heat. Bacon rejected all this. It is no accident that he was one of the formulators of the kinetic theory of heat, which construed a qualitatively experienced thermal difference as itself nothing more than the agitation of particles. Newton's theory of light is in this same anti-Scholastic tradition. He refuses to emulate late medieval attempts at 'causal' accounts of phenomena. For him it is enough to discover how to make the descriptions of phenomena intelligible. A simple undulatory theory, besides flouting the indisputable rectilinear propagation of light, would also have flouted the whole spirit of the new scientific philosophy as opposed to the old scholastic natural philosophy.

For Newton a beam of light was essentially an intricate conspiracy of particles, attended in but a secondary way by a vibratory undulation in the ether.[1]

Why then, on this theory, must light hasten through the denser medium? The writer of the *Principia* had a forceful reply. As a ray of sunlight enters (at an angle) a denser medium, like water or glass, it refracts toward the normal. This indicated to Newton an attraction between particles: the denser medium being composed of the greater material aggregate, it pulls the particles of light down out of the less dense medium through which they had been travelling, into the body of the water, or glass. So far, so good: Newton gives us a consistent atomistic picture of matter-particles attracting light-particles, from which one might indeed expect particles of light to bend into the denser matter according to Snell's Law, $\sin i / \sin r = \mu$ (where μ is the index of refraction). One consequence is obvious: if the water can thus bend the paths of light-particles as a result of attraction, it must accelerate them too, since it increases the vertical component of the velocity without decreasing the horizontal component. In other words, if water can alter a light-particle's velocity by changing its direction, then, when the new direction is established, the speed of the particle must also increase, assuming the water's 'attractive force' to remain constant throughout the transaction. Imagine the light-ray falling vertically on the refracting surface. Here, no bending occurs: however, since it is the same water whose attractive force bent an inclined beam, it must speed up a vertical beam where no force is dissipated in bending. *Ergo* on Newton's corpuscular hypothesis, light must move more quickly the denser the medium.[2] By an ingenious arrangement of rotating mirrors and fixed reflectors, Foucault[3] was able to get a good determination of the velocity of light through air. He used principles later refined by Michelson in getting a virtually absolute value. But Foucault also directed one reflected beam through water: all other factors in the experiment remained constant. The mirror's rotational speed had to be slowed down in order to get the same effect as had been observed in air; yet on Newton's hypothesis it ought to have been spun more quickly.

Now this experiment *is* crucial between the two theories, because, in the absence of special and sophisticated supplementary hypotheses about the real nature of light, Newton's theory cannot embrace Foucault's results by, for example, appealing to an inclusive sense of disjunction. No special theory is needed to see the impossibility of a ray of light both accelerating and decelerating, on entering a dense medium. So, where Newton might have embraced the findings of Young and Fresnel within his own modified corpuscular theory, he could never say that light goes both faster and slower when it enters a denser medium; he could not say this and still maintain the theory which now goes by his name.

The crucial nature of Foucault's experiment also depends on an implicit assumption; namely, that *nothing can at once accelerate and decelerate*. This differs from the assumption made by Young and Fresnel, viz. that *light must be either particulate or wave-like but not both*. While Newton can reject the latter without changing the fundamental concepts of his theory, he cannot reject the former without scrapping this theory.

The conceptual status of Newton's *Opticks* shifts, then, as it is moved from the Young–Fresnel context to the context of the Foucault experiment. Historians incline either to speak carelessly about the crucial character of the Young, Fresnel and Foucault experiments, or—in a fit of subtle sophistication—they reject all three as logically inconclusive against the Newtonian theory. The truth is that the work of Young and Fresnel is complementary to Newton's optics, a fact of which Young was markedly aware. The Foucault experiment, however, totally destroys the Newtonian corpuscular theory as it stood in the period 1704–30.

Failure to draw this distinction makes the offending historian of science little better than those nineteenth-century iconoclasts who —because of the success of the wave theory in the hands of Young, Fresnel, Foucault, Maxwell and Lorentz—came to despise Newton's corpuscular theory in particular and his optical theory in general. It makes him, again, little better than that band of twentieth-century enthusiasts of quantum physics who refuse to see in Newton's Theory of Fits anything but *the* fundamental intuition of the modern theory of radiation. None of these positions is tenable. But of the three, the historians of science are the more culpable, for they, unlike the physicists of yesterday and today, might have been expected to know better than to accept Young–Fresnel–Fizeau–Foucault as a quartet *en bloc* against Newton, or to reject the work of all four as logically inconclusive. These experiments have differing logical structures, and they bear down on different aspects of Newton's theory.[1]

E

We are now better placed for considering the logic of crucial experiments. Any experiment which purports to decide between two rival theories, I and II, must presuppose either that I is true or that II is true, but that at least I or II is true, and that in no case are both I and II true together. Proponents of both theories must accept all this if they are to recognize an experiment as crucial to the issue between them. In describing the controversy between Priestley and Lavoisier concerning combustion, phlogiston and oxygen, Professor Toulmin notes that although the latter's experiment is usually characterized as 'crucial', it cannot have been so regarded by Priestley; he would have given an interpretation of the experiment different from that of the Frenchman. Similarly, in our example, a Newtonian who could not admit that I or II obtains, but not both, could not regard the Young–Fresnel experiments as crucial against his own corpuscular theory. It takes two to 'tell the truth'; one to tell and one to listen.

What are two scientists agreed on when they consent to such a presupposition? Nothing less than a vast conceptual background; they agree to share the solid stage on which this one experiment is performed, and in terms of which it has a similar significance for both. Such agreement is surprisingly rare in science. Let any part of that conceptual stage be unfixed and the issue cannot be decided by any one performance on that platform. Priestley could not assent to the background which animated Lavoisier; and Newton, by his feeling that a rectilineal path for a spreading wave front was impossible, could not have accepted the conceptual background against which the Young–Fresnel experimentation took place.

Furthermore, a given experiment cannot be crucial for, or against, some theory unless the consequences of the theory are unambiguously deducible. This condition is clearly met in Newton's *Opticks* in so far as the Foucault experiment bears on it, for there is no doubt whatever that Newton requires sunlight to accelerate as it enters water. The condition is not met at all in so far as the

Young–Fresnel experiments bear on the Newtonian theory, because nowhere in the *Opticks* are we told how the corpuscular theory would be affected were interference phenomena other than 'Newton's rings' actually observed. This is hardly remarkable when one considers that the Interference Principle is a nineteenth-century discovery of Young himself, however much Newton may hint at it.[1]

Ernst Mach, the physicist-historian who errs most in his evaluations of Newton's optical theory, turns this last point against Newton like a stern moralist. Grimaldi describes the fringes on the inside of the shadow of a hair. Newton does not mention these; he even denies their existence (*Opticks*, III, pt. I, obs. I). Mach could excuse Newton for having failed to observe them himself. But since the Englishman was certainly familiar with Grimaldi's treatise he therefore must have suppressed mention of these *inner* fringes in the interest of his own theory: or so Mach seems to reason. But if Newton could overlook the inner fringes while observing the shadow of a hair, it is equally conceivable that he might overlook that passage in Grimaldi wherein this observation is described. Although Mach is right to say that this was historically an important observation, the fact that Newton overlooked it both in the hair's shadow, and in Grimaldi's tract, is still compatible with a reputation of the highest scientific integrity.[2]

A final point about the logic of crucial experiments: one must resist the illusion of geometrical demonstration in experiments which purport to be crucial. These are often described in terms analogous to a *reductio ad absurdum* proof in Euclid,[3] as if a theory of empirical science could be closed off and formalized so that one observation could demolish or confirm an intricate chain of deductions. Were this model accurate, the theory flattened by a crucial experiment ought never to peep forth again. How then can the particulate theory of light have been resurrected, not as an ancillary appendage of modern physics—but as virtually its leading idea? How can a theory be destroyed, and then be gloriously confirmed at a later date? The answer is either: (1) that scientific theories are never finally demolished by so-called crucial experiments; or (2) that antiquated theories are never *per se* resurrected and confirmed. We may suppose that both (1) and (2) hold.

Consider (2) first: The particulate theory of light of Einstein,

Compton, Raman, Dirac and Heitler, so influential in this century, bears only an analogical resemblance to the ingenious theory set out in Newton's *Opticks*. The enthusiasm of some physicists—for instance, De Broglie, when he says: ' . . . by a bold stroke of intuition [Newton] tried to establish association between waves and corpuscles—the motion of a projectile as a propagation of a periodicity . . . '[1] is misplaced, misguided and misleading. It obscures the fact that our modern theory of radiation rests in part on the failure of the Maxwell–Lorentz theory to explain phenomena wholly unknown to Newton; it also clouds the fact that before Newton could have understood what it was that Planck, Einstein and Compton had observed, before he could even have recognized a similarity between his theory and theirs, he would have had to take a quick course in eighteenth- and nineteenth-century physics.[2] Newton the optical theorist was the towering genius of his day, deserving of better treatment than he received from nineteenth-century wave disciples.[3] But he was not omniscient: he was not trying to find answers to *our* problems; his own were difficult enough. It cannot be his theory which has been reinstated, but a totally different theory: one which stands on its own feet and on its own evidence and which, almost coincidentally, bears a family resemblance more to what Newton conjectured than to what the wave theorists entertained. Nor has anything since discovered shown Foucault's experiment to be false, or even to have been misinterpreted by those who thought it disconfirmatory of Newton's particulate hypothesis.[4]

Now consider (1) above: beyond noting that old theories never return unchanged, it might also be remarked that theories never get old because of one crucial experiment. A redoubtable Newtonian might survive the Foucault experiment, by judiciously redefining his fundamental terms.[5] But by Foucault's time a cumulative weight of other evidence was building up against the corpuscular theory. An industrious mathematician might succeed in describing the universe as geocentric: given unlimited ingenuity one could transform every statement of contemporary celestial mechanics into an operationally equivalent statement in which the fixed point in the universe is the earth's centre of gravity. But the calculations would take a lifetime and would be worthless, because, although no single observation, e.g. Galileo's detection of all the phases of Venus, by itself demolished the geocentric theory, the accumulated weight of evidence against it has become so great that by now no one interested in discovery would try to save it. Similarly, an ingenious mathematician might be able to reformulate Newton's original optical theory so as to accommodate all the eighteenth-, nineteenth- and twentieth-century evidence against it.[1] But why should one try?

F

One final speculation—a mere speculation. What might have been the development of science through the eighteenth and nineteenth centuries had the Euclidean ideal of a scientific theory, and its attendant notion of the *experimentum crucis*, not in fact carried the day?[2] How might things have gone had Newton's inclusive sense of disjunction been tolerated by the wave theorists?

Surely the astonishment of twentieth-century physicists over the discoveries of Planck, Einstein and Compton would not have been as great as it was, since there would have been little inclination in the nineteenth century to mould the concepts 'particle' and 'wave' in logical opposition to each other. The salient feature of Newton's 'logic' is that these concepts should be compatible; that a physical event can, without contradiction, share particle and wave properties. Foucault's work would probably not have led to the demolition of Newton's theory, and to its low reputation in the nineteenth century, but would have encouraged some mathematician, someone of Clerk Maxwell's powers perhaps, to redesign the Newtonian idea of a light corpuscle, and to find a comprehensive expression for the Theory of Fits which would also embrace the new discoveries of the Victorian era. That this is possible is apparent from the analogous work of De Broglie and Schrödinger in our century, work which consisted in marrying concepts at least as incompatible as anything in Newton's system. And, had Newton's flexibility prevailed, we should have had different ideas about crucial experiments. Our propensity to regard the history of science as a rectilinear series of 'all or nothing' decisions would have been steeply minimized. In this event, decisive 'all or nothing' experiment with respect to any theory (e.g. Newton's) could occur only by introducing arbitrary dogmas concerning what a scientific theory *must* be. We would surely have allowed for more latitude in our thinking about the nature of conflict between scientific theories. But in fact none of this did happen: Newton's theory of light did not prevail. We must therefore be all the more cautious in our considerations of crucial experiments. We have had to change our ideas on the nature of light. It is to be expected that our ideas of crucial experiments, historically so intimately connected with the controversy concerning light, may likewise require radical revision.

All this constitutes a first faltering step towards the positron. The conceptual preparation necessary for rendering an observation or an experiment decisive is apparent enough in the work of Young, Fresnel and Foucault. Understanding the similar preparation in ideas which preceded the discovery of the positive electron is essential to appreciating the height of the theoretical obstacles, and achievements, in the research of Dirac, Anderson and Blackett. The unspoken assumptions controlling scientific thought play as strong a logical role now as they have in the past. More generally, Newton's conception of Fits represents just that kind of theoretical complexity which identifies the positron idea; without natural philosophers having wrestled with tricky notions of the earlier kind, the positron concept would have been intolerably difficult to form.

CHAPTER II

EXPLAINING AND PREDICTING

...An explanation...is not complete unless it might as well have functioned as a prediction; if the final event can be derived from the initial conditions and universal hypotheses stated in the explanation, then it might as well have been predicted, before it actually happened, on the basis of a knowledge of the initial conditions and general laws....[1]   CARL HEMPEL

A

For example, in September 1956 Mars was closer to earth than it had been in 24 years. But besides the good view, Mars entertained us by halting beneath Pegasus. It then moved 'backward' (i.e. from east to west). After weeks of this retrograde motion, the planet again stopped and proceeded 'forward' (from west to east).

Call this whole complex phenomenon 'P'. What is the explanation of P? A natural way to explain it is to show how P follows from: (E_1) Mars' mean distances from the Sun and the Earth; (E_2) Mars' mean period of revolution; (E_3) Mars' mean angular velocity; (E_4) Mars' position on Christmas Day 1955. We must show how P follows from all this [E_1, E_2, E_3, E_4] via the laws of celestial mechanics (which include among others (L_1) Kepler's three laws; (L_2) Galileo's two laws; and (L_3) the laws of Huygens and Newton). Thus P is explained by showing how it follows from E_1, E_2, E_3, E_4, via L_1, L_2, L_3.

On this account, however, P could have been predicted before September 1956, simply from the knowledge of E_i, the initial conditions, by extrapolation via L_j—the laws. For Hempel, explaining P is predicting P after it has occurred.

Of course, predictions *per se* are true or false. Not so explanations. They are adequate or inadequate. So Hempel's thesis must be that the justification for a prediction of P is symmetrical with the explanation of P.

This is the ideal situation Hempel describes in the quotation above. The history of science, however, presents but few examples of disciplines wherein this optimum state of affairs has been realized. Aristotle's cosmology, while it did explain the perturbations of the celestial bodies, could not predict where any planet might appear at any time.[1] Still, his 'word-pictures' made the cosmos more intelligible to his contemporaries, and, in some sense, this must count as 'explanation'. To deny this is to legislate how 'explanation' ought to be used for certain philosophical purposes, and to leave undiscussed what, in the past, have actually counted as explanations. Granted Aristotle's explanation of the cosmos to have been inadequate, it was nevertheless an explanation: when a prediction turns out to be false we do not deny that it was ever a prediction at all. But while it did, in a sense, explain the planetary motions, Aristotle's heavily ensphered cosmos could not render even a false prediction. It was not made for that purpose. Is it not in this respect like the account an historian might give of, say, the decline of the British Empire? He can explain it in just that sense appropriate in explaining historical events. But he has not been trying to predict this event, or any other. So what he says, and what Aristotle says, cannot be construed as a false prediction. At most, both offer inadequate explanations.

The great astronomers of the ancient world, however—Eudoxos, Apollonios, Hipparchos, Claudius Ptolemy—could predict where planets and stars would appear at future dates. But each explicitly rules out the possibility of explaining the physics behind the apparent motions of the cosmos: theirs was the problem of forecasting where familiar points of light might later be found on the inverted black bowl of the heavens.[2] Indeed, the history of planetary theory could be viewed as a conceptual struggle between two opposed forces, the urge to explain and the urge to predict. In some men, for example, Aristotle, the need to explain dominated: their premature attempts to give philosophically coherent, intelligible pictures of what lay 'behind' the cosmos resulted in systems which, whatever their other virtues may have been, had no use for navigation, for agriculture, or for calendrical purposes. For others, for example, Ptolemy, practical considerations were paramount; indeed they were the incentive for developing reliable predictional astronomy, purged of the cosmological overtones which dominated speculative philosophy. Ptolemy sacrificed a philosophically coherent and unified picture of the heavens in favour of an accurate instrument of forecast. Through the Middle Ages—in the Merton

College astronomers, Sacrobosco, Apian, Cusa, Oresme, Regiomontanus, Copernicus and Kepler—these opposed tendencies operate.

Only in Newton's *Principia* does the ideal Hempel sketches seem fully to be realized, even though in the second part of this chapter the role of that work in Hempel's argument is shown to be complex. The *Principia* does indeed supply a single system, L, such that if we can explain any planetary perturbation, P, in terms of its symbolism and prior facts, E_1–E_4, then we might have predicted that P in terms of that same L and E_1–E_4. Mars' retrograde motion is elegantly explained and predicted in the *Principia*.

But physics and astronomy were soon knocked from this pinnacle by Leverrier. The reasoning which led to his (and, independently, Adams') spectacular prediction of the existence of Neptune[1] was employed to deal with the superficially similar aberrations in the perihelion of Mercury, a phenomenon discovered by Leverrier himself. But the Frenchman had by now acquired the 'Newtonian habit': so he explained Mercury's problems by inventing the planet Vulcan. Leverrier warned us, however, that we could rarely observe Vulcan, since either it was always obscured by rays of the sun, or it was an incredibly dense body of small dimensions, smaller than our instruments could detect, or 'it' was really a cloud of asteroids. It *must* exist since Newtonian mechanics is true; but Newtonian mechanics would not be true unless something like Vulcan were responsible for Mercury's perturbations. This kind of assumption worked perfectly in accounting for the behaviour of Uranus and Neptune, and the problem here appears the same; only here Vulcan will probably remain unobservable.

This argument did not succeed, not only because of the 'unobservability' condition, but also because on one of Leverrier's earlier suggestions Vulcan, Sun and Earth would have had to constitute a straight-line solution to the three-body problem, a solution which is demonstrably unstable. Moreover, since Vulcan, although intra-Mercurial, had to share (on another Leverrier hypothesis) the Earth's period of revolution, in order to remain behind the Sun, the hypothesis 'falsifies' Kepler's IIIrd law: $T^2 \propto r^3$. *Reductio ad absurdum.*[2] Newtonian mechanics, and physics generally, have never been the same since.

Perhaps Hempel has outlined only an ideal situation. Possibly the situation has been realized but once, briefly, in the history of science; that need not matter. But it does matter: we must investigate further. Hempel's original statement was directed to historians. He argued that a discipline like history could be put on a logical basis as sound as that of the natural sciences. Historians and philosophers of history have found much to disagree with in Hempel's thesis: but they all concede that he does complete justice to explanation and prediction in an ideal physics. Indeed, our examples support this concession.

Still, it might be argued, as many physicists, for example, Boltzmann,[1] have argued, that Newton's *Principia* was *ab initio* an unrealistically ideal system. Of course the situation Hempel describes is realized in Newton's work; but only because the phenomena Newton observed were liberally laced with unobservables such as punctiform masses, invisible forces, absolute space and time, and a host of other mathematically elegant but observationally dubious entities. Newton should properly have been discussing his data statistically, not geometrically; he should have stuck to the letter of *Hypotheses non fingo*, allowing into his physics only entities which could be observed, operated upon, or experimented with. To qualify his exposition in this way would have been, for Newton, a labour even greater than writing the *Principia* itself. But then at least his successors would have been in no doubt as to what parts of the Newtonian system are strictly connected with phenomena, and what parts are hypothetical constructions postulated to link the phenomena into a powerful Euclidean-type network of propositions. In short, every 'law' in *Principia* contains a lot more than the facts $E_1 \ldots E_n$ which suggested it. We are never clearly told how much more; we are never given reasons for Newton's unique elaboration of the data. There are, of course, perfectly plausible reasons for his elaborations. These, however, we must learn from study and inference, not simply from reading his Latin.

This was the position advanced by Boltzmann after writing his magnificent work on the theory of gases. Gas theory, kinetic theory generally, is interpretable on a Newtonian-mechanical basis only if the latter is itself interpreted statistically. Like every physicist, Boltzmann wanted one unified theory. Hence his only recourse was to re-interpret Newton as a distribution theorist, treating the laws of the *Principia* as statistical generalizations of observations rather than quasi-geometrical axioms based on an intuition of what is ontologically self-evident.

Hempel might counter here that the limitations imposed on the 'kinetic' physicist are observational limitations merely. Statistics enter only because we are incapable of dealing in simple mechanical terms with such complex phenomena as thermally excited gases *en bloc*. The principle, however, is unaffected: theoretically it remains possible to describe the perturbations of any molecule in such a gas volume. Hence, it ought to make sense, theoretically, to predict any future state of such a molecule and thus to imagine an explanation in terms of these fundamental micro-perturbations of all the large-scale phenomena we encounter in our macro-observations.

Thus, were anyone to argue against Hempel that, in gas theory, it is never possible to predict the future dispositions of a gas's constituent molecules, it could be countered that this limitation is due only to the paucity of our information and the poverty of our computational powers. It is a technical, contingent limitation; it in no way alters our concepts. Experimental thermodynamics thus reveals no impossibility in the idea of having a complete specification of the internal constitution of a hot gas at any time. If there is no impossibility in this, there is no impossibility in the idea of predicting future micro-dispositions of such a gas, or in explaining observed macro-states of that gas in terms of its internal micro-geometry.

The situation is, however, totally different in quantum physics. Here the facts run against the schemata suggested by Hempel. True, given any single quantum phenomenon[1] P it can be completely explained *ex post facto*. One can understand fully just what kind of event occurred, in terms of the well established laws of the composite quantum theory of Jordan, Dirac, Heisenberg, and the later developments of Dyson, Schwinger and Tomonaga. These laws give the meaning of 'explaining single micro-events'. Philosophers of science should not legislate here: they must note what *counts* as explanation in microphysics, and then describe its logic precisely. It is, of course, the most fundamental feature of these quantum laws that the prediction of such a P is, as a matter of principle, impossible. This impossibility is not comparable with what obtains in classical statistical mechanics. There, we could not predict, because we did not have all the data: still, it made sense to speak of having all the data [cf. La Place]. Here, with quantum phenomena, we cannot predict, because we cannot possibly have 'all the data': there exists no conception of what it would be like to have data beyond those with which any well designed quantum mechanical problem does begin. The only theory which explains, in any sense of 'explain', quantum phenomena has built into it as a notational rule the impossibility of predicting such singular phenomena. To try to forecast, à la Hempel, the exact future co-ordinates of a high-speed electron would require manipulating the operators of the algebraic formalism so as to violate the logic governing their use. The syntax of Dirac's quantum theory contains restrictions against predicting such a phenomenon, and so to attempt the prediction would be to move outside the quantum theory, back into classical physics, which explains practically no microphysical observations.

Perhaps the discovery that, although we can explain P, we cannot, as a matter of principle, predict it, reveals an inadequacy in quantum theory itself. This has been the view of some illustrious scientists, e.g. Einstein, De Broglie and Schrödinger. But not one of these thinkers has met an obligation which any opponent of the existing theories faces.

What precisely is the concept we are asked to entertain when invited to imagine a quantum theory fundamentally different from what in fact it now is? What picture is being painted for us by the man who suggests that, although our techniques of observation make it impossible to specify the position of a high-speed electron, it none the less makes sense to think of that electron as having a position beneath the limits of present observation?[1] What exactly will microphysics be like, not in detail but in its broad structure, when within it the complete predictability of electronic future states becomes possible? Current theory could not even describe such a phenomenon: it would be a totally different theory if it could. (This is what the celebrated von Neumann 'proof of the impossibility of hidden variables' establishes.)

The answer to these rhetorical queries is that we are being asked to entertain what is theoretically untenable, empirically false, or logically meaningless. There just is no concept in quantum physics corresponding to the 'classical' ideal of Einstein, Schrödinger,

De Broglie, Bohm, Vigier (and Hempel?). Why is this point put so strongly here? There is now one, and only one, physical theory which deals at all successfully with the perplexing phenomena which, at the beginning of this century, upset classical physics. Black body radiation, the photoelectric effect, Brownian movement, X-ray diffraction, the Compton effect, electronic, atomic, and molecular diffraction, the Zeeman effect: classical theory patently cannot explain these phenomena, quantum theory can. It is, however, an essential logical feature of the new theory that it be formulated within a non-commutative algebra. This means that the position and momentum operators of the theory are of the form, $pm - mp = n$: $(n \neq 0)$. This is no superficial blemish within the system: it is its logical structure. Indeed, von Neumann once argued that all of quantum mechanics can be generated from a suitable observational interpretation of this non-commutativity formula alone. Two things immediately entailed by the acceptance of this formula are (1) there is no intelligible way in this system of speaking of the exact position of an electron of precisely known energy,[1] and (2) there is no way in this system of forming predictions of certain types of events, e.g. single neutron emission from unstable carbon. Yet this being the only way we have of discussing these phenomena at all—this theory providing us with the only consistent and empirically applicable concepts of elementary particles there are—the invitation to treat it as untenable is equivalent to an invitation to treat all current explanations of the phenomena cited as essentially untenable.

Now, no microphysicist will deny that there is a lot more to be done with respect to all the phenomena mentioned. Formal difficulties with 'renormalization' and the recent jolt given by Yang and Lee spring to mind. Still, few would deny that we are on roughly the right lines: a quantum-theoretic account of these phenomena does explain, to a considerable extent, what kind of physical events they are. What is the physicist being asked to do by the man who says there is something wrong with the quantum theory, but is totally unable to suggest a detailed alternative?[2] Objectors rarely even point out what is wrong with the current theory, other than that it fails to resemble the classical theory in obvious respects. The Hempelian symmetry between explanation and prediction is, I suspect, a consequence of classical determinism. In this respect, then, contemporary non-deterministic theories fail to support Hempel's thesis. The situation resembles what astronomers encountered in the mid nineteenth century. It was becoming clear that the celestial mechanics of Newton's *Principia*, and La Place's *Mécanique céleste*, was failing. But so long as no alternative theory was suggested for dealing with complex phenomena such as Mercury's aberrations, astronomers did not abandon the imperfect concepts they already had. This would have been a counsel of submission to ignorance, an instruction to stop thinking altogether about planetary motion. Similarly, today's critics of microphysics have presented no clear idea of what we are to use as conceptual machinery in elementary particle theory, once we have followed their advice and relegated quantum mechanics to the status of a recognizably inadequate first approximation to some future, but as-yet-undiscerned, set of ideas about microphysical nature. What are we to think? What are we to think with?

These are not merely historical observations. Should someone claim he has a good reason for abandoning a theory θ, but can suggest no alternative to θ, no other way to form concepts about the phenomena θ covers, I deny that he has good reason for abandoning θ! This, even when θ is inconsistent in spots.

All this has implications for Hempel's philosophical message. He says we never really have an explanation of P unless we could, by imagining a temporal shift, have predicted P by the same 'explanatory' techniques; explaining P is simply 'predicting' P after it has happened. But if this is meant to describe what is actually done in science, where are examples of Hempel's thesis in action? The answer is always the same: Newtonian mechanics.

That quantum theory is *fundamentally* non-deterministic is made apparent as follows: by the Principle of Identity, every negative electron must be indistinguishable, save for its co-ordinates, from every other negative electron. The same is true of every positive electron, every proton and anti-proton, neutron, anti-neutron, pion, muon—and so forth. To doubt this principle is to doubt the possibility of micro-physical science altogether—for it simply lays down the requirement that there should be one basic wave equation for every type of particle, not for every individual particle within the same type. The principle is therefore as much a part of classical physics as of modern physics. The law of freely falling bodies does not vary from body to body and fall to fall: in the same way there can be nothing, save position, to distinguish the nuclei of a cluster of atoms of carbon$_{14}$. At any one time, all C_{14} nuclei are absolutely identical: they all have the same properties, to the same degree. But C_{14}, being an unstable isotope of carbon, decays randomly, and fundamental nucleons are emitted from it in a wholly unpredictable way, as the Uncertainty Relations require. It is empirically not the case that all the nuclei of a C_{14} cluster decay at once: but then any particular nucleonic decay must be an 'uncaused' event, for, if all the nuclei are identical until one of them cracks up, then there can be no causal reason for the decay. There is nothing to distinguish the context in which the event did occur from neighbouring contexts in which it did not, save the occurrence itself.[1] It is interesting to speculate on what Hempel's analysis of explanation and prediction might have been like had this been his paradigm example, instead of, as one suspects, those of classical mechanics.

It appears that there is an intimate connexion between Hempel's 'symmetry between explanation and prediction', and the logic of Newton's *Principia*. To have learned this from Hempel is to have learned something important—about explanation and prediction, and about Newtonian science. But the professional historian need not regard himself as a Newtonian scientist; the quantum physicist necessarily cannot do so. There may be more to be said about the logic of explanation and prediction, as these concepts obtain in fields other than that one in which Hempel's analysis appears sound. Philosophers might still wish to know what is the logical structure of explanation and prediction as these ideas function in living, growing sciences, and not only as they were employed for a brief period in a discipline which is by now little more than a computing device for rocket and missile engineers.

B

I wish now to amend the foregoing somewhat by shifting the spot-light to Hempel's account of prediction. If he is in any way correct in stressing a symmetry between reasons supporting an explanation of x and reasons supporting a prediction of that same x, then we might have expected as much discussion of his account of prediction as has centred on his analysis of explanation.

My argument turns on the historical point which seems to have influenced Hempel so much. Contrast an eighteenth-century reaction to Newton's *Principia* with an early-twentieth-century reaction to that same theory. After 1687 Leibniz and Poleni were critical: to them *Principia* seemed merely a mathematical predicting device. Leibniz felt that this theory related to its observational data precisely as did Ptolemy's *Almagest* to its data. As we saw, Ptolemy stressed that he could never hope to *explain* the wanderings of the planets. His aspiration was only to find a geometrical calculus through which he might forecast when next a given planet would halt in its eastward motion, 'back up' a few degrees, and then continue forward again. Understanding beyond this exceeded Ptolemy's objectives. The *Almagest* was but a computational machine; it had some success in saying when celestial events might occur, but did not even undertake to ask why they occurred. Leibniz contended that the *Principia* had the same epistemic function *vis-à-vis* the data of mechanics. By providing a network of formulae, Newton enabled us to predict with precision the future motions of celestial bodies, the behaviour of projectiles, the tides, and so forth. These predictions are anchored to the law of universal gravitation and the three laws of motion which bear Newton's name. Such laws provide a formal framework for generating numbers describing future events, but they are themselves neither explained nor explicable. Newton's sentiment is: 'It is enough to have provided a formula by which mechanical behaviour can be predicted. Beyond this, "explaining" gravity, or space, or time, or inertia...etc., does not concern me.' For natural philosophers like Leibniz, however, such an attitude could not be reconciled with the idea of a complete science. Hence, throughout the early eighteenth century Newton's physics seemed only to predict—never to explain. It illustrated the 'black box' conception of a scientific theory, of which we have heard so much. Feed the data numbers into the box, turn the handle [of the theory], and let the prediction numbers tumble out.

Contrast this view of Newton's mechanics with the one current in the nineteenth century—as encapsulated in C. D. Broad's book of 1913, *Perception, Physics, and Reality*. Broad describes Newton's laws as constituting the paradigms of causal explanation. After demonstrating a formal connexion between mechanical phenomena and these laws, no further understanding was required, since this very demonstration constituted the most complete explanation possible.

In two centuries, therefore, a change had occurred which was of great philosophical importance. Yet the theory in question, Newton's mechanics, remained essentially the same: the refinements of La Place, Lagrange, Maxwell and Hertz did not affect its formal structure, although it did make it function more smoothly as a piece of inferential machinery.

But now, since the theory was identically structured at both times, Hempel's thesis, if correct, should obtain with equal force in either context. Whatever counted as a prediction of x in the eighteenth century should then have been logically symmetrical with some corresponding explanation of x: what counted as a prediction of x in the twentieth century should also be symmetrical with a corresponding explanation. But the conceptual difference between Leibniz and Broad is not illuminated by Hempel's analysis. Leibniz would have characterized Newton's mechanics as grinding out a mere prediction of x, in the total absence of a corresponding explanation. Broad felt no such reservation: the same network of inferences which failed to convey understanding of mechanical phenomena to Leibniz constituted (for Broad) everything which 'understanding mechanical phenomena' could mean. Since Hempel is concerned only with networks of inference *per se*, his account cannot resolve this historical difference.

Our purpose now is to explore the notion of mere predictability. This may help us analyse the attitudes of Leibniz and Broad, and also to discover where Hempel's thesis really obtains. What counts as mere predictability (i.e. numerical forecasting) at one time may serve at another time as full-blown prediction (i.e. advance explanation of future events). Hempel cannot distinguish the two: hence he has been attacked for appearing to give an account of the latter, when perhaps his thesis has concerned the former.

Let us think ourselves back into the heyday of classical mechanics. This would be about 1847—just after Leverrier's brilliant prediction of the existence of Neptune by an extrapolation of the explanatory techniques of Newton's *Principia*. (Adams made the same prediction independently.) This triumph raised the theory to the highest pinnacle it ever had, or ever has, known. One or two minor local flaws marred the complete victory of the theory.

Imagine that at such a time a young mathematician, let us call him 'Notwen', comes forward with 'an alternative to Newton's theory', invoked initially just to deal with those minor flaws. The alternative is algebraically complex. To use Notwenian theory at all requires skill in unfamiliar branches of functional analysis. Moreover, the fundamental postulates of Notwen's theory are replete with uninterpreted terms like $\sqrt{-1}$, 'negative velocities', 'infinite densities', etc. But suppose also that this new theory, put forward in 1847, turned out results identical to those which orthodox Newtonian mechanics could then achieve, as well as coping with those minor flaws which started Notwen working in the first place. (This is the first stage in our historical thought experiment.)

In short, when first announced, the new and unfamiliar Notwenian theory seems to generate all the predictions and all the numbers that ordinary Newtonian theory can generate, and also patches up some minor flaws. What, then, would be the standard attitude towards this 'alternative'? I submit that it would be regarded as a mathematical curiosity, and that there would be considerable puzzlement concerning how and why it works at all. It would be described as a mere predicting device—an intra-mathematical analogue of a mysteriously complex machine which, as if by numerological magic, seems to bring forth the correct answers to all questions. In such a context no one would try to *explain* phenomena by appeal to Notwen's system.

Suppose further that orthodox Newtonian mechanics begins to show major weaknesses. Problems it cannot solve, and events it cannot predict, turn up with increasing frequency—while at the same time Notwen's theory, by simple extension, succeeds with such new problems and events with remarkable precision. This alone will decide nothing; but scientists will begin to show increasing reliance on the new 'alternative to Newton'. Courses dealing with the newer mathematical techniques necessary within Notwenian mechanics will come to be taught in the better universities. The 'old guard' will insist that the Notwen theory gains its accuracy 'by accident'—and explains nothing. But the energies of

younger physicists will turn more and more to the internal development of Notwenian physics. (This is the second stage in our
imaginary example.)

Suppose, finally, that as Newtonian mechanics continues to fall
apart, Notwenian physics opens up new branches of science,
focuses on problems never before perceived, fuses disciplines
thought before to be distinct, and sharpens experimental techniques
to an unprecedented degree. Imagine the whole scientific enterprise caught up within the basic metabolism of Notwenian physics.
The very pattern of thinking within any inquiry properly called
'scientific' will reflect that of the new physics—which has by now
become virtually synonymous with the concepts of 'science' and
'scientific explanation'. To be able to cope with a scientific problem
at all will just be to have become able to build it into the conceptual
framework of the Notwenian physics. (This is the third stage in
our *gedankenexperiment*.)

Consider these three stages more schematically:

(1) First comes the presentation of an algorithmic novelty.
Some intricate piece of formalism is introduced which, miracu-
lously, churns out all the observational consequences of some
older and more familiar theory, as well as patching up some minor
flaws in the latter. This is what is sometimes called a 'black box'
theory—no deep understanding about phenomena follows directly
from the successful use of the algorithm. Still, scientists seem
prepared to use the new technique on trial because of its capacity
to get numerical results. Then they translate these results into the
more familiar terms of the orthodox theory in order to provide
understanding.

(2) The next stage consists in the new formalism beginning to
outstrip the extant theory with respect to 'predictive power'.
Although affording no more 'insight' into the phenomena than
before, the new theory appears now not simply as a remarkable
algebraic alternative to the orthodox theory; it has become essential
for getting numbers at the predictional-observational level. It will
have ceased being merely a 'black box'—it is now a 'grey box'.
It is still opaque so far as providing understanding of phenomena
goes, but it appears that there must be some fundamental reason
why the new formalism works in predictions where the old theory
has collapsed. The new theory is no longer viewed merely as

mathematical magic; it now seems that its structure, and that of the phenomena with which it deals, must have something in common. Otherwise it could not be succeeding where the older, more familiar, theory had failed.

(3) At this stage, the new theory will have pushed into fields far from the minor phenomena with which it began. It will connect subject-matters formerly thought distinct. The new theory will have so permeated the operations and techniques of the body of science that its structure will appear to be *the* pattern of a scientific inquiry. At this stage it will seem a 'glass box'. Because of the pervasive patterns of reasoning which the new theory establishes within so many related disciplines, scientists will cease distinguishing between its structure and that of the phenomena themselves. The equations of the new theory will seem actually to mirror the processes of nature; the presuppositions of the theory will constitute fountainheads of understanding for everything flowing from those suppositions. And indeed, what else *can* one think about phenomena other than what the currently most successful physical theory permits one to think? In short, the theory will have become known by its fruits—its capacity to provide understanding will have grown in direct ratio to its capacity to generate successful predictions within increasingly wider areas of inquiry. And our very idea of what 'understanding' means will have grown, and changed, with the growth and changes of the theory. So also will our idea of 'explanation'.

These three stages characterize the development of classical mechanics itself. When Newton enunciated his 'mathematical philosophy', there were several prose-laden theories of nature already in the field: the theory of impetus, Kepler's celestial 'spokes of force', Cartesian vortex mechanics, and others. These theories purported to make phenomena intelligible, by relating them to intuitively evident first principles, in a manner which Newton's contemporaries felt he had not achieved. Thus Leibniz's attack on Newton's mechanics is a first-stage assault. It is much like what would have been a typical reaction to Notwenian mechanics in 1847. Broad's adulation is typical of stage three; by 1900 the Newtonian pattern of thinking had permeated every corner of the house of science. The place that Newtonian first principles occupied within an immense pattern of interlocking analyses and types of inquiry made it seem like a glass-box theory when C. D. Broad wrote. In a profound sense the theory seemed to provide a glimpse of the innermost workings of nature itself. It was as if '$F \propto \gamma Mm/r^2$' in some way *pictured* something in nature. But for Leibniz these extensive and manifold ramifications of the theory in other extra-mechanical disciplines were absent, so in his day it was construed as being merely a 'black box'.

It is tempting to characterize the theory of quantum mechanics as being already well advanced into its second phase—its grey-box stage. For many years it has been providing answers to questions about microphysical nature which Newtonian theory is incapable of answering. Moreover, it seems no longer to be regarded as merely constituting the mathematical conjury which dominated the period between 1913 and 1927. Still, it has not yet permeated enough into related fields, such as theoretical chemistry or genetics, to make many insist that it constitutes a paradigm of scientific explanation; on the contrary, many philosophers mark deficiencies in the explanatory framework of quantum mechanics. This may signal only a delayed stage-one reaction, or it may reflect the fact that the stage-two successes of Quantum Mechanics have not been as spectacular as were those of Newtonian Mechanics in the late eighteenth and early nineteenth centuries.

In any case, that Leibniz did not, while Broad did, feel Newtonian mechanics to constitute the best of all possible explanations of nature, reflects a clear conceptual difference between them. This difference is not simply a function of internal changes within the theory: the basic inferential patterns of classical mechanics have undergone no substantial modifications from the seventeenth century up to the present time. The difference between them, however, may be a function of the differences between the degree to which Newtonian mechanics had permeated and interlocked with every scientific discipline by 1900, as contrasted with the relative absence of any such systematic and synoptic effect in the *Principia* in 1700—at least so far as disciplines outside mechanics (strictly interpreted) were concerned. The wider conceptual 'set' of a physical theory is not merely a matter of psychological importance. It affects our understanding of the conceptual status of a given theory, of its logical relationship to other theories and to observations—and to our ideas concerning what it is to understand something, or to explain it. Just as we know something about an animate object when we see how its insides function, so we know more about it when we also see how the object as a whole functions in relation to other organisms, of the same kind, and of different kinds. With physical theories it is much the same. When born, they are subjected to minute micro-analysis; but at this stage it is rarely clear what they may ultimately be able to *do*. When what they can do becomes known, even some of the earlier micro-analysis may change. Leibniz did not know how much the *Principia* was going to be able to do, and Broad did. Their different attitudes towards its explanatory powers are a reflexion of this difference.

Hempel's account of the logical symmetry between explanation and prediction is sound as a piece of conceptual micro-analysis. The logical structure of theories is such that, for every well made prediction, there will be some correspondingly well made postdiction.[1] This is as true of every well made seventeenth-century prediction as it is of every well made twentieth-century prediction. However, not every postdiction will count at all times as an explanation. A postdiction embedded in a far-flung system of scientific theories *may* count as an explanation. At another time, however, the same pattern of inference, the same postdiction, may constitute nothing of the kind.

By 1913 Newton's postdictions had become explanations: Newtonian theory had become a glass box. 'Backward inferences' within the theory (i.e. inferences from present or past events to their known initial conditions) seemed like backward glances at nature's inner workings. For Leibniz, however, the postdictions were *merely* postdictions: they were simply arguments reversed along the time parameter. The theory was a black box. Backward inferences were for Leibniz no more than the consideration back-to-front of a sequence of mathematical moves involving *t*. Predictions in the *Principia*, therefore, were *mere* predictions for Leibniz; they were logically symmetrical only with postdictions. For Broad, predictions within Newtonian theory were mature; they were logically symmetrical with explanations—i.e. postdictions seen now as parts of an immense interlocking pattern of related hypotheses, inferences and observational data. Hempel's analysis *per se* cannot distinguish these two cases—cannot distinguish prediction from *mere* prediction. But the entire history of science is representable as much as a relentless advance from mere prediction to prediction as it is an advance from postdiction to explanation. Both prediction and explanation are concepts reflecting networks and patterns of theories: postdiction and *mere* prediction reflect only the internal structure and direction of particular inferences in particular sciences. By generalizing a logical truth of the latter kind, Hempel may have misled us concerning a conceptual issue of the former kind.

Here, then, is a second step towards understanding the positive electron. An inflexible application of Hempel's logical criteria would make it extraordinarily difficult to appreciate in what sense Dirac predicted a positron in 1931; for he was 'backed into' having to do so by an otherwise successful theory which possessed this prediction as an awkward, but logically indispensable, consequence. Hempel's unrestrained thesis would tax our capacity to understand in what sense the 'hole theory' of the positive electron constituted an explanation of the observations made by Millikan, Anderson, Blackett and Occhialini. The 'hole theory' as such never served to *predict* the existence of any positron. The positron story is very complex: in some ways it has an idea-structure *sui generis*. We must see what 'explanation' and 'prediction' mean within that structure, rather than press classically orientated conceptions into place no matter what the effect on adjacent parts of the structure.

However, §B of this chapter has attempted to locate the real force of Hempel's argument: that in a well made theory for every *mere* prediction there must be a corresponding postdiction (in the sense described). This is certainly true of Dirac's 1928 paper, and of his further studies on the positron before 1931. Only when he sought a physical interpretation of his 'negative energy' solutions—in order to turn theoretical postdictions into explanations, and mere predictions into predictions—did the conceptual battle for the positive electron begin.

CHAPTER III

PICTURING

A[1]

Physicists exhort us not to try to picture atomic particles. This can be puzzling, for how can one discover, and interfere with, unpicturable, unvisualizable objects? And with what instruments? How can such entities be conceived at all? There are certain properties which atomic particles must necessarily lack: electrons could not be other than in principle unpicturable. The impossibility of visualizing ultimate matter is an essential feature of atomic explanation.

Suppose you requested an explanation of the properties of chlorine gas: its green colour and memorable odour. Would the following satisfy you? 'The peculiar colour and odour of chlorine derive from this: the gas is composed of many tiny units, each one of which has the colour and the odour in question.'[2]

Would this be adequate? Many physicists would doubt whether it was an explanation at all. Those who explained cohesion between bodies by inventing hooked atoms were accused by Newton of 'begging the question'.[3] Seeber denied any brick-like structure in crystals; this would have required investing the bricks with just those properties of crystals requiring explanation.[4] Similar accounts were advanced by Clerk Maxwell,[5] von Laue[6] and Dirac:[7] all these thinkers noted and rejected 'explanations' of this kind, since they do not answer questions about material properties, but only postpone them. What requires explanation cannot itself figure in the explanation; as we saw in chapter I, we would not be satisfied were the sleep-inducing qualities of opium explained by reference to its soporific properties. The explanation has merely been deferred.

One might object: the dynamical behaviour of a billiard ball can be explained by the behaviour of some other ball which has just struck it; one could explain why a cloud moves by referring to the motions of its constituent molecules, whose 'group' motion is the cloud's motion; to say that blood is made up of red particles in some sense explains the redness of blood.

True enough. But one cannot explain why any one thing is red by saying that all red things contain red particles; nor could one explain why any single thing moves by noting that all moving things contain moving particles. In general, though each member of a class of events may be explained by other members, the totality of the class cannot be explained by any member of the class. The totality of red things cannot be explained by anything which is red; the totality of movement cannot be explained by anything which moves. Finally, all the picturable properties of objects, the totality of them, cannot be explained by reference to anything which itself possesses any of those properties.

The history of atomism serves as another illustration. The Greek Natural Philosophers[1] sought to explain the immense diversity of physical properties. Thales, Empedocles and Democritus agreed that the myriad colours, odours, tastes and textures of things were not each one final and irreducible, but could be analysed yet further: they were the manifestations of something more fundamental. Most nominations for this 'more fundamental something' failed because they possessed the properties to be explained. Water could not be just a liquid, not just a vapour (fog), not just solid (ice). Were it basically any one of these, e.g. liquid, how could reference to it explain the solid and vaporous things we observe every day? On the other hand, if 'water' named a trinity of types of matter—a 'liquid-vapour–solid'—any explanation of material properties in terms of it would be complex and mysterious. Were such properties as these abandoned, however, why should the fundamental substance be called 'water' at all?

The concepts of earth, water, air and fire constituted Empedocles' attempt to meet this logical difficulty: by the theoretical blending of these idealized elements he sought to explain all properties of objects. But this reasoning was inelegant and uneconomical, and it left the ideas of solidity, liquescence, vaporousness and heat themselves unexplained.

Democritus saw that if this 'fundamental something' was to explain all the observed properties of objects, it could not itself possess any of those properties. Earth, water, air and fire did possess them: Democritus' atoms therefore lacked all properties, save only geometrical and dynamical ones. All such atoms were identical and purged of 'secondary qualities'. 'An object merely

43

appears to have colour; it merely appears to be sweet or bitter. Only atoms and empty space have a real existence.'[1] This already renders the atom unpicturable. Can a colourless atom be pictured? Windows and spectacles can be pictured only because at certain angles they are not transparent.[2] If the colours of objects are to be explained by atoms, then atoms cannot be coloured, nor can they be pictured.

The original request to explain the properties of chlorine was not a query about a local phenomenon, this particular bottle of gas with these special properties. What was sought was a general theoretical account of the properties of chlorine, such that it affects us as it does. Merely to endow atoms of the gas with these same optical and chemical properties requiring explanation is to refuse to supply such a theoretical account.

What is it to supply a theory? It is at least this: to offer an intelligible, systematic, conceptual pattern for the observed data. The value of such a pattern lies in its capacity to unite phenomena which, without the theory, are either surprising, anomalous, or left wholly unnoticed.

Democritus' atomic theory avoids investing atoms with secondary properties which themselves require explanation. It provides a pattern of concepts in virtue of which the properties the atom does possess—position, shape, motion—can account for the other, secondary properties of objects. The conceptual price paid for this intellectual gain is unpicturability—and unpicturability *in principle* at that.

Atomic explanation did not change through the centuries following. Later scholars were trapped within 'stage (1)' of atomic theory, as discussed earlier on p. 37; they were unable to visualize atoms, just as were Democritus' contemporaries. As the theory moved through 'stage (2)' into (3), and gained support in Chemistry and Physics, however, scientists came to regard atoms as almost familiar things.[3] When speaking strictly[4] they renounced the picturable atom. But why speak so strictly? The geometer never denies himself the use of drawn lines. These should be one-dimensional; but geometrical constructions cannot be carried out with one-dimensional lines. Similarly, physicists could think about atoms only by visualizing them; and why not if it helped them to secure explanations? Thus the near-invisible diagrams of geometry crept into physical thinking about atoms. Rutherford proceeded along these lines in 1911 when he accepted Nagaoka's idea 'of a "Saturnian" atom which...consist(s) of a central attracting mass surrounded by rings of rotating electrons'.[1] Atoms should have been as unpicturable as the entities of geometry: but no physicist chose so to paralyse his thinking. In fact, the atoms became the very models of geometrical and dynamical behaviour. This made them eminently picturable. Why should colours and lines have to be more than merely a practical necessity? Like the perfect circles of ancient geometry the classical atoms were the ultimate limit of a series of sketches of increasing fineness.[2]

Even this expedient can no longer serve the physicist's imagination. Atomic explanation has always ruled out secondary qualities such as colours, odours and tastes; but modern atomic explanation even denies its fundamental units any direct correspondence with the primary qualities, the traditional dimensions, positions and dynamical properties. In classical physics kinematical studies precede dynamical ones: in quantum physics this division and order is hardly feasible. Primary qualities were fundamental to the statical-kinematical conceptions which classical particle theory chose to build into a Euclidean space, and dynamical properties of bodies were ancillary to these. But now the order is reversed: an atomic particle's statical-kinematical properties are determined by its dynamical properties—quantum dynamics is the prior discipline. The basic concept of microphysics is interaction.

The Democritean–Newtonian–Daltonian atom simply cannot explain what has been observed during this century; its postulated properties—impenetrability, homogeneity, sphericity—can no longer pattern and integrate our data.[3] To account for all the now known facts the atom must be a complex system of more fundamental entities.[4] Electrons, protons, neutrons, mesons, antiprotons, anti-neutrons, X-ray photons and hyperons have been detected. Others are likely to be discovered too if certain 'gaps' in our experiments are to be explicable; but these cannot, in any sense, be the point-particles of classical natural philosophy.[5]

The properties of particles are discovered and, in a way, determined by the physicist. Certain phenomena are observed which are surprising and require explanation: the observations may be of the tracks left by microparticles in a cloud chamber, or in a photographic emulsion, or they may be the scintillations excited when particles strike certain sensitive screens, or any one of a number of their other indirect effects.[1] The theoretician seeks concepts from which he can generate explanations of the phenomena. From the properties which he ascribes to atomic entities he hopes to infer to what has been encountered in the laboratory; he aspires to fix the data into an intelligible conceptual pattern. When this is achieved he will know what properties fundamental entities do have.

For example, electrons scintillate, and 'veer away' from negatively charged matter, so they must be somewhat like particles. But electron beams also diffract like beams of light, so they must resemble waves too.[2] In order to explain such phenomena as these, the physicist must fashion his concept of the electron so as to facilitate inferences both to its particle and to its wave behaviour; and a conception so fashioned is unavoidably unpicturable.

Observations may multiply. Further properties may be pushed back into the concept 'electron', properties from which each new explanation of each new observation follows as a consequence. That theory which depends on the particle being assumed to have these properties will naturally be taken to explain the observations: unless, of course, it leads to unsound inferences in other directions. At this point one could have no reason to doubt the real existence of the properties which intelligibility demands of these subatomic entities.[3] The result is the most radical unpicturability. If microphysical explanation is even to begin, it must as a matter of logic presuppose theoretical entities endowed with exactly such a delicate, and wholly non-classical, cluster of properties.[4]

In general, if A, B and C can be explained only by assuming some other phenomenon to have properties α, β and γ, then this is the *best possible reason* for taking this other phenomenon to possess α, β and γ.[5]

In macrophysics, such an hypothesis is tested by looking at this 'other phenomenon' to see if it has α, β and γ. With elementary particles, however, we cannot simply look. All we have to go on are the large-scale phenomena A, B and C (ionization tracks, bubble-trails, scintillations, etc.) and perhaps future phenomena D, E and F. Hence one must suppose that the particles actually have the 'explanatory' properties in question, α, β and γ, and see if, by mathematical manipulation of these, we can infer to further theoretical properties δ, ϵ and ϕ, which might explain the further phenomena D, E and F.

The cluster of properties, α, β and γ, may constitute an unpicturable conceptual entity to begin with. As new properties δ, ϵ and ϕ are 'worked into' our idea of the particle[1] the unpicturability can become profound. This does not matter: there will never be any atomic particles we will fail to recognize just because we failed to form an identification picture of them in advance. The main point about fundamental particles is that they show themselves to have just those properties which they must have if they are to explain the larger-scale phenomena requiring explanation. Thus, discovering the properties of elementary particles consists in a logical complex which is, in principle, like the very one within which Democritus found himself. Unless they are taken to have certain abstract properties the elementary particles cannot explain the phenomena they were invoked to explain.

Professor Fermi illustrated this: 'The existence of the neutrino has been suggested...as an alternative to the apparent lack of conservation of energy in beta disintegrations. It is neutral. Its mass appears to be either zero or extremely small....Its spin is believed to be $\frac{1}{2}$; its magnetic moment either zero or very small....'[2] Our concepts of the properties of the neutrino are determined by there being gross phenomena A, B and C, which defy explanation unless an entity exists having the properties α, β and γ: just those which the neutrino has. The idea of the neutrino, like those of other atomic particles, is a conceptual construction 'backwards' from what we observe in the large. The principles which guarantee the neutrino's existence are like those which guarantee the existence of electrons, α particles, and even atoms. This does not make the subject-matter of atomic physics any less real—elementary particles are not logical fictions; they are not mathematically divined hypotheses spirited from the theoretician's imagination to serve as bases for his deductions. Nor does knowledge of elementary particles consist merely in a summary description of what we learn directly through large-scale observation. What we must realize is that knowledge of this portion of the micro-world is derived by means far more complex than such philosophically easy accounts suggest.

Again, the situation is as follows: a surprising phenomenon is observed. We expect the energy released by homogeneous radio-active substances to depend solely on the initial and end stages of the nucleus (hence all α rays ejected from a homogeneous substance have the same range, i.e. the same energy). But β particles are emitted with all possible energies (Chadwick): and this apparently contradicts the Principle of Conservation of Energy.

Accept the hypothesis (of Pauli); with each β-particle another particle also leaves the nucleus, carrying the difference in energy. Suppose this particle is construed (following Fermi) as having properties: velocity c, hence mass = 0, and in no case greater than $\frac{1}{500}$th of an electron mass [recently lowered by Langer *et al.* to a maximum of $\frac{1}{2000}$th of an electron mass]; charge neutral; magnetic moment = 0, or very small. If we accept all this, then the continuous spectrum of β-ray decay will be explicable, and the energy principle still holds.

Yes, but why accept this concept of the neutrino? It cannot be observed in a Wilson chamber or a bubble chamber—nor has it ever been directly detected by another means, prior to the effect discovered in 1956 by Cowan and Reines. Besides, such a particle seems both unlikely and unsettling. So why accept the neutrino?

Because if you do, the continuous β-ray spectrum will be explained, and the energy principle will remain intact. What, indeed, could be a better reason?[1]

The formation of the neutrino concept provides a paradigm of how observation and theory—physics and mathematics—have been laced together in physical explanation. Mathematical techniques more subtle and powerful than anything within the geometry of Kepler, Galileo, Beeckman, Descartes and Newton are vital to today's physical thinking. Only these highly sophisticated algebraic techniques can organize into one broad system of explanation the chaotically diverse and unpicturable properties which fundamental particles must have if observed phenomena are to be explained at all. As Heisenberg puts it, '...the totality of Schrödinger's differential equations corresponds to the totality of all possible states of atoms and chemical compounds'. He even dreams of '...a single equation from which will follow the properties of matter in general'.[2]

Concerning mental pictures, the present situation in fundamental physics could not have been much different from what it is. We are now faced with unpicturability-in-principle: to try to picture particles at all is to rob oneself of what is needed to explain the properties of ordinary physical objects.[1] Though intrinsically unpicturable and unimaginable, these mathematically described particles can explain the behaviour of matter in the most powerful manner known in the history of physics. Indeed, only when the quest for picturability in physics ended was the essence of explanation within all Natural Philosophy laid bare.

B

William Whewell wrote in 1833:

...If we in our thoughts attempt to divest matter of its powers of resisting and moving, it ceases to be matter, according to our conceptions, and we can no longer reason upon it with any distinctness. And yet...the properties of matter...do not obtain by any absolute necessity....[2]

Within the subsequent century the matter-concept underwent radical changes. Let us explore these changes and note how they relate to the distinction between primary and secondary properties.

Determining the essence of the matter-concept was, as we have just seen, a problem already familiar to Democritus, and to Galileo. Locke gives the distinction its classic shape:

The qualities then that are in bodies, rightly considered, are...: First, the bulk, figure, number, situation, and motion or rest of their solid parts. Those are in them, whether we perceive them or not; and when they are of that size that we can discover them, we have by these an idea of the thing as it is in itself; as is plain in artificial things. These I call *primary qualities*. Secondly, the power that is in any body, by reason of its insensible primary qualities, to operate after a peculiar manner any of our senses, and thereby produce in *us* the different ideas of several colours, sounds, smells, tastes, etc. These are usually called *sensible qualities*....The first of these...may be properly called real, original, or primary qualities; because they are in the things themselves, whether they are perceived or not: and upon their different modifications it is that the secondary qualities depend.[3]

Thus on the one hand there are the properties matter *really has*; these are geometrical, statical and dynamical. Shape, mass, motion and impact—these are a body's primary properties. However, its apparent colour in ultraviolet light—or daylight—its taste, its tone, its fragrance, are the body's secondary properties. Our appreciation of these latter varies with the state of our senses; secondary properties result from interaction between percipient and perceived. But the primary qualities seem to be 'in the bodies themselves'; hence, they are the very properties of matter itself. As historians know, the primary-secondary distinction dissolved in George Berkeley's inkwell. Knowing a body's shape seemed to the bishop as much the result of interaction as any secondary property. [Eighteenth-century psychologists knew that a given mass could generate variable perceptions in subjects differently conditioned.] Berkeley's epistemology, therefore, erased any distinction in principle between primary and secondary properties. Either the two were equally weak, or equally strong—depending on how one interprets Berkeley. Either secondary properties are just as basic to matter as the primaries, or the primaries give no more indication of matter 'as it really is' than the secondaries. The latter seems more like Berkeley; hence I adopt it here.

Berkeley's analyses, however, seemed *merely* philosophical. Distinctions between primaries and secondaries may indeed fail under strict analysis; but Berkeley's scientific contemporaries still treated the distinction as fundamental, philosophers notwithstanding. Scientists were concerned with the physical properties of objects, not their 'real' properties. We shall return to this distinction.

Consider now Heisenberg's insight into the history of atomism and the manner in which it reflects the primary-secondary contrast: 'It is impossible to explain... qualities of matter except by tracing these back to the behaviour of entities which themselves no longer possess these qualities. If atoms are really to explain the origin of colour and smell of visible material bodies, then they cannot possess properties like colour and smell...Atomic theory consistently denies the atom any such perceptible qualities.'[1]

Boyle made a similar point: 'Matter being in its own nature but one, the diversity we see in bodies must necessarily arise from somewhat else than the matter they consist of.'[2] Lucretius' atoms were colourless; an aggregate's colour depended on the size, shape and interrelations of its constituent atoms.[3] His atoms were without heat, sound, taste or smell.[4] And Bacon wrote 'Bodies entirely even in the particles which affect vision are transparent, bodies simply uneven are white, bodies uneven and in a compound yet regular texture are all colours except black; while bodies uneven and in a compound, irregular, and confused texture are black.'[5]

Birch writes of Newton: 'The atoms...were themselves, he thought, transparent; opacity was caused by "the multitude of reflections caused in their internal parts".'[1]

Thus the atomic hypothesis, and its intricate history, would crumble unless the ancient distinction between primary and secondary qualities braced it. No classical atomist thought atoms to be coloured, fragrant, hot, or tastable; the basic function of atoms was to explain away such properties as but the molar manifestations of the atom's primary properties and geometrical configurations.

Not every atomist stressed the same atomic primary, although all agreed that, whatever they were, the atom's properties were necessarily primary, an argument to which we shall return. The atom's primaries usually included properties such as position, shape and motion.

Position was paramount for Democritus, but it was *shape* for Epicurus and Lucretius. Newton fixed on the *motions* of atoms. Gassendi remarked their *combinatory properties*; this already constitutes an extension of the Lockean notion. But doubtless combinatory capacity would have been accepted by all as a primary property, although not every atomist would have stressed it *à la* Gassendi. Henceforth, the term 'primary' will be used in this extended way. Atomic *irresolvability* attracted Boyle, but this is clearly tautological. For Lavoisier, Richter and Dalton, *mass* was basic. Berzelius stressed their *binding force* [again this falls within our extended class of primaries]. Further properties were stressed by Faraday, Weber, Maxwell, Boltzmann, Clausius, Mayer, Loschmidt and Hittorf. But, by all, the atoms were characterized by some cluster of primary properties, on a selected one of which further theoretical constructions were founded. The exception is Stumpf, who could not imagine atoms as spatial bodies lacking colour;[2] but he is the exception proving the rule—by which is meant 'probing the rule': we know what is generally true when we note how a counterinstance deviates.

The predominant sentiment of the Scientific Revolution was expressed by Newton: 'I...suspect that [the phenomena of nature] may all depend upon certain forces by which the particles of bodies...are either mutually impelled towards one another and cohere in regular figures, or are repelled and recede from one another.'[1] Here is a yet wider extension of Locke's use of 'primary'. But forces which impel and repel would surely be on the primary side of the ancient fence.

The degree to which the primary-secondary distinction remained scientifically fundamental, despite Berkeley's levelling analysis, is illustrated by Euler: 'The whole of natural science consists in showing in what state the bodies were when this or that change took place, and that...just that change had to take place which actually occurred.'[2] Helmholtz is as direct: 'The task of physical science is to reduce all phenomena of nature to forces of attraction and repulsion the intensity of which is dependent only upon the mutual distance of material bodies. Only if this problem is solved are we sure that nature is conceivable.'[3]

These sentiments reflect a spectrum of related attitudes: the mechanical philosophy, theoretical determinism, the reduction of all science to physics. These are generable only from an implicit atomism. Historically, this devolves into something resembling the classical distinction between primary and secondary properties.

'Resembling' is an operative word. Berkeley had speculated about the *real* properties of matter. He felt the classical primary-secondary distinction to be unsound. These properties were on the *same* epistemic level so far as knowing 'reality' was concerned. One of the bishop's scientific contemporaries could grant this, however, and yet preserve the same distinction at a different level—that concerned not with matter's *real* properties (a philosopher's inquiry at most), but with its *physical* properties (a scientist's inquiry at least).

Berkeley's 'Thou shalt not speak of primary properties as philosophically real' is hence distinguishable from a prohibition heard in this century: 'Thou shalt not speak of primary properties as physically real'. In the eighteenth century a scientist could have accepted the first and rejected the second. Now he may very well accept the second, whatever may be his attitude towards the first.

Hence, so far as one was concerned with distinguishing primary properties (which were real and in matter itself) from secondary properties (which were merely produced in us)—Berkeley's epistemic criticism was devastating. Still, the distinction remained viable in natural philosophy, the province not of philosophically real, but of physically real properties. Physically real properties contrast with mere appearances (intersubjectively understood). That a submerged stick is really straight contrasts with its bent appearance, and that the solidified CO_2 is really cold contrasts with our impression of it as blistering hot. But concern with the philosophically real undercuts these scientific inquiries altogether. The latter concern stable, permanent properties of objects as contrasted with those which are evanescent and contextually dependent, those which accrue to them via special conditions of observation (e.g. ultraviolet illumination, or intoxicated observers). Berkeley's inquiry is concerned with the philosophical extension of this scientific contrast, as signalled by the question: 'And which of these kinds of properties does matter *really have?*' The scientist's use of the primary-secondary distinction is restricted to delineating the contrast 'observed under all conditions' against 'observed only under special conditions'. The philosopher asks the more pervasive question, which Berkeley answers by a denial: there are no better grounds for supposing matter really has properties we regularly observe it to have, than there are for thinking its properties are those we irregularly observe.

Within the class of physically real properties scientists did distinguish primaries from secondaries—they had to do so to sustain an intelligible atomism. But Berkeley's objectives were not scientific; he dismissed the distinction as philosophically untenable. Physical science is only now undergoing its Berkeleyan self-criticism; when Whewell wrote, it had not done so. He spoke of conceptual constraints against tinkering with our ideas of the primary physical properties of matter. These constraints are no longer as binding as in the nineteenth century; and indeed, our understanding of elementary matter, of electrons, cannot proceed within a classical conception of primary physical properties.

For a theory of electrons to succeed now, the electron-idea must be divested of its *classical* conceptions of resisting and moving. This is what Whewell claimed we could not do: 'Divest matter of its powers of resisting and moving...and we can no longer reason upon it with any distinctness.' But electrons can be reasoned upon with distinctness, although, it may be granted, they remain unfamiliar material objects.

C

Let us now consider the representative answers to 'classical' questions about the electron, that most fundamental of particles. What is the 'diameter' of an electron? The usual theoretical answer is that it is of the order of 6×10^{-13} cm. Experimentally this is not determinable: such a magnitude would be very difficult to detect because of pion and nucleon pair-creation phenomena at the required energies (80 BeV). The concept of electronic diameter is based on the formula $d = 2e - 2e^2/m_0 c^2 \approx 5 \cdot 7 \times 10^{-13}$ cm; or, $r \cong e^2/m_0 c^2 \approx 2 \cdot 81785 \times 10^{-13}$ cm[1]; no evidence contradicts this, but the fine determination is beyond current laboratory technique. Of course, some theoreticians put the diameter at 0, arguing that this is compatible with electron-proton scattering experiments at 1 BeV. But notice that the quantum electrodynamics of small distances is already at stake in this question.

If theory just tolerates the electron having a diameter, it ought also to have a shape. What shape? Is it spherical? Punctiform? There is at present no experimental information enabling one to form any consistent geometrical model of the electron. The electron's charge is assumed to have spherical distribution, as with its magnetic moment; none the less, experimentalists are often prone to treat electrons as points. Again, this issue, like the 'diametral' one above, awaits such tests as the very high energy electron-electron scattering experiments now underway at Stanford.

What about the electron's 'solidity'? Again, neither theory nor experiments help. Some theoreticians feel the concept to be meaningless; others remark that the deeper electron penetration proceeds, the more difficult a decision becomes, because of the myriad new particles created by the probing particle and the target-electron. Still others think the particle may have a 'solid' central core where some current theories break down (10^{-13} cm). If this is not the case, then the electron can only be described as a cloud of virtual particles plus a central bare point charge.

Other magnitudes within electron physics are readily determinable. Collisions are understood: there are sound theories within quantum electrodynamics and myriad experiments on electron scattering properties. The electron's rest mass (after 'renormalization') is determinable: $m_0 = 9 \cdot 1083 \times 10^{-28}$ g, a quantity confirmed in many divergent types of experiment. But again, the relation between electronic mass and charge is troublesome. (Attempts to understand mass in self-energy terms have been unsuccessful.) The spin-angular-momentum of the electron is always $\frac{1}{2}$; this is, again, well established by the Zeeman effect, and other experiments: and quantal transformations of the electron (as theoretically represented in the Lorentz group) require precisely this spin. The rest energy of the electron is $m_0 c^2$, as disclosed in electron-positron pair production: this energy value determines the development of the electron's state in time. There are no *de facto* negative energies encountered in electron physics. However, negative frequencies make theoretical sense.

All this must be appreciated lest it seem that science's total knowledge of the electron is too slight to permit generalizations about today's matter-concept. A great deal is known about the particle, but what is known seems incompatible with classical ideas about matter. Thus, while one can always speak of the state of a classical particle, i.e. its simultaneous position and velocity, nothing like this can even be articulated in quantum theory, wherein the position and momentum operators are managed according to the rule: $XP - PX = (h/2\pi i)$. This has implications: is it that physics is just not yet in a position to determine electron states? No. Quantum mechanics is the only theory through which electrons can now be understood at all; that theory excludes the very possibility of forming a consistent concept of an electronic particle's state.[1] Either we speak precisely of its position, or of its momentum, but not precisely of both at once. Similarly, we can speak of a person then as a bachelor, and now as married. But we cannot speak of him as being at once married and a bachelor; the reason is analogous to that within microphysics. Conceptual tension results in either case. This is not to say that the tension in both cases is identical: it is not.

It is often mooted that the conceptual pain of indeterminancy is restricted to the physics of very tiny and very brief phenomena. But it has in fact macrophysical consequences: a Geiger counter intercepting β particles from an unstable isotopic source will click in a wholly unpredictable way. Once a particle has been emitted, the counter's click is determined classically. But it remains conceptually untenable to predict when a particle *will* be emitted, and hence when the counter will next click. This is a logical feature of the only available means for understanding intra-atomic phenomena. Indeed, a Geiger counter so arranged constitutes the perfect randomizer. Its macrophysical clicks must be in principle unpredictable. There is no comprehensible alternative.

Still further classical properties of matter are jolted in electron theory. Electron solidity was just described as a concept for which there is no relevant quantum theory or experiment; 'being in contact with an electron' has the same null status, although some theoreticians feel sympathetic to the possibility, especially when this is interpreted in field-theoretic terms. A spectacular departure from classical theory is the process of particle-creation, first described in the early 1930's. That particles could 'materialize' out of radiation is an idea for which twentieth-century physics had no preparation. Joliot and Curie, Millikan and Anderson, Blackett and Occhialini, Fermi and Uhlenbeck, noted oppositely curving cloud-chamber tracks of identical range emanating from a common point within a radiative source: and this was a new phenomenon. The mass-energy equivalence had long been known; still, it remained implicit in molar physics that while matter could be transformed from this shape to that, or from one state to another, it could never become other than matter: nor could it be created from other than matter. The discovery of the positron crushed this assumption; matter (e.g. electrons) can be created out of energy alone. Yet, when electrons are created, one cannot speak of their states, shapes, or solidity in the familiar molar ways (the full details will be explored in chapter IX). Furthermore, the theoretical reason for supposing that two electrons could not simultaneously occupy the same place is not overpoweringly strong, even though the Pauli exclusion principle has never been experimentally violated. And when an orbital electron is excited in the H-atom, it jumps out to a wider orbit; yet one has no way of speaking of it as having ever been between the orbits. There is no workable concept of an electron's age—unless it be taken as ∞—nor any intelligible conception of its density. Again, this is not ignorance comparable with the limitations of our knowledge concerning Venus. In the latter case, we lack facts; but we know what it would be like to have them. Within electron theory, the limitations referred to are built into the conceptual structure of the theory itself. We do not know now what it would be like to manage electrons as we do, save in terms of the theories and concepts we have actually got. Change the concepts and you change our current theories: but *until* the theories are changed, we must do as they now instruct us to do, namely, abandon earlier notions of the properties of material particles. (These points are elaborated in more detail in appendix II.)

Matter has been dematerialized, not just as a concept of the philosophically real, but now as an idea of modern physics. Matter can be analysed down to the level of fundamental particles; but at that depth the direction of the analysis changes, and this constitutes a major conceptual surprise in the history of science. The things which for Newton typified matter—e.g. an exactly determinable state, a point shape, absolute solidity—these are now the properties electrons do not, because theoretically they cannot, have.

In other words, modern science has dematerialized matter more radically than Berkeley did. He showed that, despite ancient epistemic dogmas, primary and secondary properties were in the same conceptual boat. One of his scientific contemporaries could have inferred from the bishop's analyses that primary properties were just as weak as the secondaries as indicators of the real properties of matter: he could have concluded this and still continued to do consistent physics. For Berkeley's criticism was abstractly philosophical; it concerned our knowledge of the 'real' properties of matter, as opposed to its physical properties; it left Newtonian mechanics intact and unscathed. [Similarly, perplexities of contemporary epistemology have no effect on today's mechanical engineers.]

The dematerialization of matter encountered in this century, however, has rocked mechanics to its foundations. As an intraphysical revolution in ideas, this compares with the intra-mathematical revolution initiated by Gödel. Some scientists still think of electrons as point-masses with most of the properties of minute billiard balls—just as some mathematicians still have Formalist (i.e. Hilbertian) hankerings. But how unclear such physicists can be when questioned about the nature of things like β-beam interference patterns. Either they say nothing at all, or nothing at all intelligible (usually capped with some remark like 'I am an empiricist'). In the eighteenth century one could accept Berkeley's demonstration of the inadequacy of primary properties as indicators of 'real' matter, and still do consistent physics; much as today a psychologist can grant that there are philosophical problems about other minds, and still rely on the verbal responses of his subjects. But the twentieth-century's dematerialization of matter has made it conceptually impossible to accept a Newtonian picture of the properties of matter and still do consistent physics.

Some will assent to much I have said here, and yet will qualify my conclusion. They may grant that the ancient distinction between primary and secondary properties, like philosophy itself, branched into the natural philosophy of the seventeenth century, and the 'pure' philosophy of the eighteenth century—the latter as typified in Berkeley. An intimate historical connexion between the successful growth of atomism in science and the correlative dependence of scientists on some version of the primary-secondary distinction might also be granted. Perhaps it will even be conceded that Berkeley's challenge to this distinction affected only the epistemological branch of the conceptual tree, not its scientific branch. The latter has been affected only by *contemporary* matter theory, wherein any correspondence between the properties which matter (e.g. electrons) is now known to have, and the classical 'primary' properties, is at best analogical, and at worst non-existent.

These are my theses thus far. From them, however, some will not conclude, as I do, that modern physics has destroyed our intra-scientific version of the primary-secondary distinction, rather as Berkeley destroyed its intra-philosophical version. At least one critic will torment the body of my argument by hacking off its tail, as follows:

Granted, the properties electrons are now known to have, α, β, γ, δ, ..., may be different from the properties classically termed 'primary'. This does not destroy the primary-secondary distinction: quite the contrary. For if the electron *has* α, β, γ, δ, ..., then, however dissimilar from the classical primaries of philosophy and natural philosophy, then α, β, γ, δ, ..., *are* (along with the properties of other particles) the primary properties of matter, whatever they may be. And these primary electronic (protonic, neutronic) properties contrast with other manifestations of elementary particles and their aggregates, which disclose themselves only through interactions between observers and things observed, e.g. manifestations like the colours, tastes and odours of macrophysical objects (which are, after all, but constellations of fundamental particles). Granted, physics has changed the values appropriate for the property-variables α, β, γ, δ, ..., still, the primary-secondary distinction remains viable so long as there are good reasons for claiming that fundamental particles *do* have α, β, γ, δ, ..., etc. It remains viable so long as some properties of aggregates, and some properties of components-of-aggregates, are distinguishable in that the former result from observer-interaction whereas the latter, however unfamiliar, are such that we have good theoretical reasons for thinking them observer-independent. A theory of the electron is a theory about the properties electrons *have*, not a theory describing what bubbles up out of electron-observer interactions. The primary-secondary distinction of classical physics has now become a contrast between the objectifiable and the non-objectifiable properties of microparticles.

This specific criticism gets airborne only via a runway of concessions to my general thesis, to establish the plausibility of which has been my only objective here. I disagree with the entire spirit of this criticism and its hankering after complete objectifiability within quantum theory; I will indicate why in chapter v. But there is in this criticism no challenge to the historical point that *our* ideas about the primary properties of particles are different from those of the tradition concerned with primary properties (despite Whewell's contention that no such change could occur, cf. p. 49). Nor have I perceived here any challenge to the further point, that Berkeley's attack on the primary-secondary distinction left physicists free to exploit the distinction in their atomistic theories of the eighteenth and nineteenth centuries, whereas they are no longer free to do this in the old way. Since these main points are unaffected by the contention that the objectifiability-non-objectifiability contrast is the same as primary-secondary contrast (with the property-values left unspecified), I will back off now with only the remark that even this contention may be demonstrated to be indefensible.

This has been a third long step towards the positron concept. Since our capacities to picture microparticles and to invest them with primary properties have suffered limitations, an understanding of the positron will not result from trying to picture it in Locke's or Whewell's terms. (Indeed, even the 'hole' theory of the positive electron is to some degree parasitic on such picturability; to that degree it is not helpful.) The stature of the positron discovery increases, when contrasted with discoveries of other objects easily pictured as having a classical 'state', mass, shape, and other primary properties.

59

CHAPTER IV

CORRESPONDENCE AND UNCERTAINTY

Our references to microphysics, and to quantum theory, have so far been unsystematic and selective; from this point forward the pace will quicken and our coverage will be more thorough. In this chapter a qualitative analysis of the relations between macrophysics and microphysics may carry us further towards the positron concept.

Quantum physics contains classical physics as a limiting case[1]

Does it? In a very limited, highly technical sense, yes. Still, the Correspondence Principle and the Uncertainty Principle are inevitably in conceptual conflict; and the latter, being a basic part of the logic of quantum physics, almost always triumphs.

But how can a system structured by one logic 'contain' another system, one characterized by a fundamentally different logic, incompatible with the first? Part of the answer may lie in observing that microphysics and macrophysics are now, and always have been, two independent systems. Their parts are exactly analogous where they 'overlap', i.e. where one is free to regard the H-atom as either a very small classical body, or as a very large quantum body. Here, the relevant equations in the one system may be symbolically identical with those of the other; none the less they remain equations in different systems. Their logical structure remains distinct, despite the identity of their symbolic form.[2]

The Correspondence Principle of quantum physics, on one interpretation, must be at tension with the Uncertainty Principle. Weyl says: 'Thus we see a new quantum physics emerge of which the old classical laws are a limiting case, in the same sense as Einstein's relativistic mechanic passes into Newton's mechanic when c, the velocity of light, tends to ∞'.[3]

This is now a familiar pronouncement. Treatises in theoretical physics intend something special when they so describe the Correspondence Principle. In such a context one is rarely misled;[4] in other contexts, however, misconceptions can arise.

Weyl's words could, for instance, lead one into the following perplexity:

(a) The motion of a planet, e.g. Mars, is described and explained in terms of 'the old classical laws'. These descriptions proceed as follows: in practice one cannot determine a planet's state by absolutely 'sharp' co-ordinates and momentum vectors; still, it is always correct and intelligible to speak of it *as having* exact co-ordinates and momentum. In classical mechanics uncertainties in state determination are in principle eradicable. Standard expositions regularly refer to punctiform masses, the paradigms of mechanical behaviour; point-particles are conceptual possibilities within classical particle physics.

(b) Elementary particle physics constitutes a different logical situation. The discoveries of 1900–30, if they were to be explained at all, forced physicists to combine concepts in unprecedented ways, e.g. $\lambda = h/mv$. A direct consequence of these combinations of concepts is expressed in $\Delta\chi.\Delta v \cong \hbar/m$, where $\Delta\chi$ and Δv measure the uncertainty in a particle's co-determined position and velocity. Within quantum theory, to speak of the *exact* co-ordinates *and* momentum of a particle at t makes no intelligible assertion at all. What could it assert? That a Schrödinger wave packet has been compressed to a geometrical point? This cannot even be false; one must at least have a clear idea of x to be able to use it in making a false statement. Is there a clear concept of a wave packet at a point? To say that there is not, is not simply to reiterate the truism that our instruments are too blunt for the delicate observations needed in order to determine the simultaneous positions and momenta of microparticles. In the well established languages of quantum theory a description of the exact 'state' of a fundamental particle cannot even be formulated, much less used in experiment. It is, to take an example, a condition of Dirac's theory that position and momentum operators are non-commutative: to let them commute is not to express anything in Dirac's theory.[1] Whatever the wave equation $(\Delta\psi + 8\pi^2 m/h^2(E - U)\psi = 0)$ can be said to express, it cannot be 'squeezed' to a geometrical point: at least, not without phase velocities spreading over all possible values. Nor can momentum be specified by a unique number without the positional co-ordinates being 'smeared' through all space. So if the Schrödinger equation is conceptually fundamental to the language of quantum theory (which it seems difficult to deny), then, for anything which could be described by the ψ function, nothing even remotely like $v = dr/dt = \dot{r}$, or

$$a = dv/dt = \dot{v} = d^2r/dt^2 = \ddot{r},$$

can obtain.[1] Point-particles, therefore, are *not* conceptual possibilities within elementary particle physics. However,

(c) Quantum theory embraces classical particle physics. '. . .we see a new quantum physics emerge, of which the old classical laws are a limiting case. . . .'[2] The justification offered for this is usually as follows: The orbital frequency of the electron in a hydrogen atom is given by $\omega/2\pi = \gamma_{(cl)} = 4\pi^2me^4/h^3n^3$. According to the classical connexion between radiation and electrical oscillation, this is the same as the radiated frequency. But quantum theory gives

$$\gamma_{(qu)} = (2\pi^2e^4m/h^3) \times (n_i^2 - n_f^2/n_i^2n_f^2)$$

for the frequency of radiation connected with the transition $n_i \to n_f$. If $n_i \to n_f$ is small compared with n_i, we can write instead

$$\gamma_{(qu)} = (4\pi^2e^4m/h^3n_i^3) \times (n_i - n_f).$$

Thus, in the limiting case of large quantum numbers, $\Delta n = 1$ gives a frequency identical with the classical frequency, i.e. $\gamma_{(qu)} = \gamma_{(cl)}$. The transition $\Delta n = 2$ gives the first harmonic $2\gamma_{(cl)}$..., and so on.

(d) It is precisely here that the perplexity arises. A certain cluster of symbols, S, is taken to express an intelligible assertion in classical mechanics; yet that same symbol-cluster S may not be so regarded in quantum mechanics.

Could $(d^2r/dt^2)m = F$ be translated into Dirac's notation? Or von Neumann's? No, not directly. None the less the languages of the systems are reputed to be logically continuous. As the Law of Inertia is said to be only a special case of the Second Law of Motion, so classical mechanics *in toto* is said to be but a special case of quantum mechanics. These are apparently just distinguishable clusters of statements within the same overall language.

Formal statements and languages do not work in this way, however. A well formed sentence, S, if it can make an intelligible empirical assertion anywhere within a formal language, must be capable of doing so everywhere within that same language. Technical notations are usually defined and delimited in terms of rules determining which symbol-combinations can be used to make intelligible assertions. If a given sentence, S, can express an intelligible statement in one context, but not in another, it would be natural to conclude that the formal languages involved in these different contexts were different formal languages. Finite versus transfinite arithmetics, Euclidean versus non-Euclidean geometries, the language of time versus the language of space, the language of mind versus that of brain; all reveal themselves as different and discontinuous on this principle. What can be said meaningfully in one case may express nothing intelligible in the other.

This also happens when S expresses the state of a particle, as in classical physics—where the result is an intelligible assertion—versus the S which purports to express the 'state' of a particle in quantum physics (e.g. in Dirac's notation)—the result being no assertion in that language at all.[1] Ordinarily this would be conclusive evidence that the languages are formally different, and logically discontinuous. But the Correspondence Principle apparently instructs us to regard them otherwise: quantum theory, as Weyl said, embraces the old classical laws as a limiting case.[2]

There is the conceptual perplexity: for how can intelligible empirical assertions within a formal language L become unintelligible within that same L just because quantum numbers get smaller? Conversely, how can unintelligible clusters of symbols within one discriminable branch of L become meaningful just because quantum numbers get larger? The intelligibility of assertions within a formal language cannot be managed in this way.[3] A spectrum of 'intelligible assertability', through which a single formula S can roam within a language, is unthinkable. Either S can make an intelligible empirical assertion in all of the language in which it figures—or else the latter is really more than one formal language.

Either the Uncertainty Principle holds, i.e. the S of classical physics makes no assertion in quantum physics, or the Correspondence Principle holds, i.e. the S of classical physics is a limiting case of quantum physics. But not both. Or else we are misinterpreting one, or both, of the Principles.[4]

First we are warned that the new physics is logically different from the old, and that we should not make old-fashioned demands on it. Then we are told that the two are conceptually quite harmonious. This needs sorting out.[5]

THE CONCEPT OF THE POSITRON

The difficulty can be expressed in terms of probability distributions. Classical theory permits the simultaneous increase of joint probabilities of accuracy (in determining parametric pairs like time-energy and position-momentum); in quantum theory this is illegitimate. But, apparently, as quantum numbers get larger, the legitimacy of these joint probabilities seems to increase: the same perplexity arises.[1]

(e) So the alternatives seem to be: (1) quantum physics cannot really embrace classical physics as a limiting case (a conclusion requiring radical re-interpretation of the Correspondence Principle); or (2) quantum physics ought not to be considered as permanently and logically restricted, allowing no analogue for the classical 'state' S within it; or (3) classical physics itself should be restricted with regard to the Uncertainty Principle construction of S, just as in quantum physics.

Alternative (3) may be dismissed. Granted, it may describe more faithfully the limitations of actual observation and experiment, as Boltzmann once remarked when arguing that classical statistical mechanics should be the foundation for all physics. Cf. p. 28. But (3) constitutes a self-denying ordinance of no practical scientific value: a classical mechanics without punctiform masses would be too difficult, conceptually and pedagogically, to justify any such a recommended change.

Alternative (2) has been adopted by several eminent physicists, mathematicians and logicians; Einstein, Rosen, De Broglie, Bohm, Moyal, Bartlett, Vigier, Popper, Jeffreys and Feyerabend, to name a few. Thus '...the limitations expressed by the leaders of quantum theory are not essential to the theory and arise simply because the theory has not yet been expressed in a sufficiently general form'.[2] The adoption of this alternative results from noting the conceptual tension between the Correspondence Principle and the Uncertainty Principle. When one then considers further the example concerning the orbital frequency of the electron in the hydrogen atom—plus thirty years of intellectual uneasiness caused by the Uncertainty Principle—alternative (2) begins to look plausible and attractive. None the less the alternative comprises a misconception as to the nature of the Correspondence Principle. We will sketch an argument here which will be developed in extenso in the next chapter.

The Uncertainty Relations are an intrinsic feature of quantum theory. They are built into $\lambda = h/mv$ and

$$\Delta\psi + 8\pi^2 m/h^2 (E - U)\,\psi = 0.$$

These relations were already implicit in the very first decisions of De Broglie (1923–24) and Schrödinger (1925–26) to weld particle and wave notions into a single algorithm. Nor is Sir Harold Jeffreys' contention (above) clear: how *exactly* could any mathematical generalization change the relationship between two logically discontinuous ideas?[1]

There is *no* ultimate logical connexion between the languages of classical physics and quantum physics—any more than there is one between a sense-datum language and a material object language. I cannot support this claim by appealing to $\Delta\chi.\Delta v \cong \hbar/m$ itself; that would be an obvious *petitio principii*. But consider the following:[2]

Classical physics is a particle dynamics set in a 'Euclidean' space-time framework. Its order of development is always Kinematics → Dynamics: one's first area of study is Galilean reference frames, vector analysis, and the properties of bodies at rest; only after such inquiries are specifically dynamical ideas introduced into this geometrical-kinematical framework. Naturally, within such studies points are just the massy intersections of one-dimensional co-ordinates—punctiform bodies. They are Euclidean points, moving in time, endowed with mass. Given any two of them, most of classical dynamics can be worked out; with any three of them arise some of the most complex computational problems in physics. There is not even now a *general* solution to the three-body problem as stated by Newton in the seventeenth century.

If anything, development in Quantum Theory reverses this order—though actually, no such division is even possible. Assuming it were, however, an elementary particle's 'kinematical' properties would then depend on its dynamical properties, and not vice versa as in classical physics. The Nagaoka, Rutherford and Bohr conceptions of the atom broke down at just this juncture: these physicists tried to work new dynamical properties into the traditional framework; and the subsequent difficulties are well known.[3]

De Broglie noted that to give a velocity c to a particle of mass > 0

would require an infinite amount of energy. He asked, however, whether such particles might be related to a wave mechanism somewhat as photons are related to the wave nature of light. Here is the first starting-point of a new pattern of ideas: a wave motion at a geometrical point is inconceivable; hence photons and electrons must 'spread' and can never be punctiform.[1]

The punctiform mass, primarily a kinematical conception, is the starting-point of classical particle theory. The wave pulse, primarily a dynamical conception, is the springboard of quantum theory. Languages leaping up from such different platforms are likely to perpetuate this logical difference throughout their development and subsequent structure; and this is indeed the case.

What then about the hydrogen atom with large quantum numbers? What is the explanation of this 'classical' result? This has been misunderstood too. Languages having such different conceptual frameworks cannot simply mesh as the algebra suggests they do; their logical gears are not compatible. Identically structured sentences and formulae, though they can express many different statements, even different types of statement, cannot express single statements whose sense and intelligibility varies with the size of quantum numbers: not unless they are really set in different languages and managed by different rules, i.e. are different statements.[2] Propositions get their force from the entire language system within which they figure. That $(4\pi^2 e^4 m / h^3 n_i^3) \times (n_i - n_f)$ gives a 'classical' frequency for the transition $\Delta n = 1$ proves at most that there is a formal analogy between certain reaches of quantum theory and certain reaches of classical theory: that it is no more than an analogy is obscured only because the same symbols are used in both languages. This no more proves a logical identity between the two than does the use of ' + ' and ' − ' for both valence theory and number theory show the latter theories to have an identical logic.

Let me give a tangential 'philosophical' illustration of this point. Men are made of cells. It might be urged that whereas one can assert that men have brains, personalities and financial worries, it is no assertion at all to say such things of cells. This would be incorrect: to say such things of cells would be intelligible, but false. Suppose, however, that 'cell-talk' were constructed so as to be *logically* different from 'man-talk': the two idioms could then never merge. 'It has schizophrenia and an overdraft' would then express no intelligible assertion at all in cell-language, just as '...is divisible by o' expresses no intelligible assertion in arithmetic. Even though a certain complex combination of cells could be spoken of in ways analogous to our manner of speaking of a man, this would not fuse the two languages: not even when both idioms are used to characterize the same physical object—me. Should one individual speak of me as a man, but another speak of me as a collection of cells, though the *denotatum* of both discourses be identical, the speakers will yet diverge conceptually. The two will not be speaking the same language.[1]

Similarly, in a sufficiently intricate sense-datum language it might be possible to construct sentences analogous to material-object sentences. That is, if in the same conditions it were true to assert a certain material object claim, S, it would also be true to assert its sense-datum analogon S'. Thus if, when it were true to say S, 'There is a bear before me', it would also be true to say S', 'I am aware of a brownish, grizzloid, ursoid patch', then the two sentences would be epistemically analogous. This will not, of course, prove the *identity* of S and S', and their associated language-systems. 'There is a bear before me' could be false even when stated sincerely. But could this be the case with 'I am aware of a brownish, grizzloid, ursoid patch'? The ranges of these two languages overlap considerably, but this no more 'reduces' one to the other than does the language of mind (memory, sensation, character, habits, imagination, personality, etc.) simply reduce to the language of the brain (synapses, neurons, cortices, lobes, etc.). Nor does the language of Picasso reduce to that of Heisenberg, even when they both speak truly, in their special ways, of a sunset. Their public utterances may be identical—'It is red now'—but their assertions diverge widely. Similarly: 'The probability that an α is a β is 1' is analogous, but not equivalent to 'All α's are β's'.[2] The conceptual differences in the language systems are not minimized by the fact that such analogous utterances can often convey truth in the same context.

The logical continuity suggested by careless semi-popular statements of the Correspondence Principle (and supported by examples of the energy-levels-of-hydrogen type) is illusory. The Principle does not show Classical Particle Physics to be a logically limiting case of Elementary Particle Physics—although, admittedly, the formalisms of these two systems may be completely analogous at many points. What the Correspondence Principle does show is that when quantum numbers are high enough the hydrogen atom can justifiably be regarded from either of two points of view: as a small macrophysical body set in classical space-time (wherein, for example, '$(d^2r/dt^2)m = F$' will serve as an intelligible form of assertion), or as a large 'quantum' body exemplifying to but a small degree the dynamics of elementary particles (where '$(d^2r/dt^2)m = F$' will not constitute the form of an intelligible assertion).[1]

In short, we are to some extent free to treat the H-atom as we please, depending on our problem. Similarly, we treat Mars and Mercury sometimes as punctiform masses, sometimes as solid oblate spheroids. We regard gases sometimes as dense, continuous media (e.g. in acoustics), and sometimes as porous, discontinuous swirls of particles (e.g. in statistical thermodynamics). A hydrogen atom *qua* small microparticle is as different conceptually from the same H-atom *qua* large microparticle as are any of the differing pairs in these examples.[2]

If one insists on some crude statement of the Correspondence Principle, then the modification necessary to relax the conceptual tension which I have described must be made in classical, not in quantum, mechanics. The electron as a point-particle in Euclidean space simply cannot explain the phenomena encountered in this century. One might, however, restrict celestial mechanics so that $\Delta\chi.\Delta v \cong \hbar/m$ (observationally, it was never entitled to 'punctiform masses' anyhow). But this restriction would make no scientific difference. Just as utterances concerning temperatures less than $-273°$ C now make no intelligible assertion within classical kinetic theory, so then utterances concerning a macro-body's exact state would be regarded as making no intelligible assertion within classical theory. This is equivalent to the recommendation that the derivative within the differential calculus be regarded as lacking any ultimate physical interpretation, which echoes the ruling that physical thinking concern itself only with observable quantities: compare the recommendation of Einstein that we abandon talk about the ether, and simultaneous inter-stellar events; compare also the recommendation of Heisenberg (as against Schrödinger) that elementary particle theorists abandon talk about individual electronic orbits, frequencies, velocities, etc., restricting itself exclusively to the scattering properties (matrices) of cathode, α and β rays—and other 'group' phenomena.

But this extreme operationalism seems unnecessarily Procrustean: classical mechanics is simpler as it is now, provided only that we remain alert to the logical properties of its notation. Once one understands the conceptual structure of a piece of discourse, there is no need to rewrite that discourse in some stilted symbolism, merely to make that structure obviously explicit. Logicians can talk science to death.

There is no logical staircase running from the physics of 10^{-28} cm to the physics of 10^{28} light years: there is at least one sharp break. That is why we can make intelligible assertions about the exact co-ordinates and momentum (i.e. the state) of Mars, but not about the elementary particles of which Mars is constituted, or even about that one elementary particle located at (or nearest to) Mars' centre of gravity. As an indication of how the mathematics of elementary particle physics can be managed, the Correspondence Principle is clear and useful.[1] Indeed, it is perfectly legitimate to point out that, in the limit of large quantum numbers, *average* properties of microparticles—position, momentum, etc.—are definable as if they were classical measures and obeyed classical equations. But when spoken of in more spectacular ways (as, for instance, by Weyl and the writers of handbooks on 'Modern Physics') the nature of intelligibility in physics hangs precariously in the balance.

This constitutes our longest stride towards the positron so far. Noting how microphysics and macrophysics are actually connected is basic to understanding how the observations of Anderson were related to the theorizing of Dirac. It has been suggested here that since macrophysics and microphysics have distinct logical structures, any apparent continuity between the two—such as that proposed by the Correspondence Principle—can at most constitute a *continuity of choice* concerning which theory is more useful and tractable for a given problem. Thus cloud-chamber experimentalists, from Skobeltzyn to Blackett, determined the energies and sources of track-leaving particles by appeals to track-curvature ($H\rho$), range, and degree of ionization: techniques which were (and

THE CONCEPT OF THE POSITRON

are) largely 'classical' in nature. Thus did Anderson detect and isolate the positron, and thus did Blackett pin it down firmly. But all electrons manifest field, or wave-like, properties too. Laue patterns and Thomson rings are also brought about experimentally; and they too are described in largely classical terms. The elementary particle theorist must hammer these divergent aspects of electrons (negative and positive) into a unified theory, resting on a few powerful, but algebraically simple, equations. The result can hardly be expected to have a conceptual structure continuous with either of the two classical varieties of electronic energy-propagation (particulate and undulatory) which provided the experimental occasion for the theory. Small wonder that Anderson saw no connexion between his particle of 1932 and Dirac's papers of 1928–31; nor did Dirac see any such connexion for some time. Blackett, in 1933, first saw the structural analogy which revealed the Anderson particle and the Dirac 'particle' to be one and the same—the positron. By suggesting too much, the Correspondence Principle hints at a closer connexion between Anderson's macro-observations and Dirac's micro-hypotheses than could ever have existed. The gap between the two cannot be closed by prose or by principle, because the gap is a logical one. One must never lose sight of this when reconstructing the positron discovery.

CHAPTER V

INTERPRETING

It has become fashionable among philosophers of science to attack the 'Copenhagen Interpretation' of quantum theory as being unrealistic,[1] unreflective,[2] or unnecessary.[3] The present chapter may be vulnerable to the same objections; but its aim is to relocate this Copenhagen 'interpretation' in its historical and conceptual context, and to argue for the virtues of orthodox quantum theory as it now holds—algorithmic inelegancies notwithstanding. Because, even should this latter argument be unconvincing, the Copenhagen interpretation *did* explicitly control microphysical thinking between 1927 and 1933, the six years out of which sprang the concept of the Positron.

A

The theory of Niels Bohr exfoliates from seeds a century old. As we have seen, the controversy over the nature of light was analogous to our present discussions about interpreting $|\psi(q)|^2$. Grimaldi's undulatory theory, as developed by Huygens, speculated about by Hooke, and confirmed by Young, Fizeau and Foucault, encountered the opposition of the 'corpuscularians', Newton, Biot, Boscovich and LaPlace. The plot is very intricate, as chapter I indicated; but it resolves somewhere within the nineteenth century, when the work of Young and Foucault came to be regarded as decisive *against* the particulate theory.

Young's work, as we also saw in chapter I, proves only that light is wave-like, not that it is non-corpuscular. The latter follows only from assuming further that 'light is either wave-like or corpuscular (but never both at once)'. Newton would not have accepted this *proviso*. None the less, Foucault *did* crush a cornerstone of Newton's *Opticks* by proving that light-velocity decreases as medium-density increases; for he so refuted Newton's theory of particulate attraction, with which Snell's law was accounted for.[4]

We have seen how the wave theorists marked this defeat. The ideas of *particle* and *wave* were designed in logical opposition to each other. Particle dynamics and electro-dynamics (or, in general, wave dynamics) matured as mutually exclusive and incompatible theories, because of (1) the apparent conclusiveness of Foucault's 'crucial' experiment, and (2) the conviction that between them one or the other of these two theories could explain every kind of energy transfer. Yet the two theories could never apply simultaneously to the same event. A particle (as we saw on p. 10) has ideally sharp co-ordinates, is in one place at one time. No two particles can share the same place; this point is built into the very logic of Newton's talk about punctiform masses. They collide and rebound—and with a calculable energy exchange. A wave disturbance, however, essentially lacks sharp co-ordinates. It spreads boundlessly through all its undulating medium. 'Wave motion at a geometrical point' would express, for Maxwell and Newton, nothing intelligible.[1] Two waves *can* be in the same place at once (as when surf waves cross at a point); yet there is nothing in wave motion like particulate collision, impact, and recoil. (This follows from the wave-theoretic law of linear superposition.)

In the algebra of Maxwell and Lorentz, one could treat wave properties α, β, γ as the obverse of some comparable class of particulate properties $\sim\alpha$, $\sim\beta$, $\sim\gamma$. It was unthinkable that an event should be at once describable both ways: by this I mean not just unimaginable, but *notationally impossible*. In the only languages available for describing particle and wave dynamics, such a joint description would have virtually constituted a contradiction. Wave and particle ideas had now become conceptual opposites.

In this lies the kernel of the Copenhagen Interpretation: twentieth-century nature refused to live up to nineteenth-century expectations. The discontinuous emission of energy from radiant, black bodies,[2] the discovery that photo-electron energy increases with the frequency of the incident light, independently of intensity,[3] the photon theory of Einstein,[4] the Compton effect,[5] and the first confirmations of the De Broglie–Schrödinger wave theory[6] of matter by Davisson, Germer and G. P. Thomson:[7] all this suggested that microparticles must be described in particulate and wave-like terms simultaneously. Yet the only such terms available for the combined description were the inflexible legacy of Maxwell's successors, terms which had been designed and structured to rule out such a simultaneous description. From the necessity of describing nature thus arise all the conceptual constraints of quantum theory—including the Copenhagen Interpretation.

In microphysics it is arbitrary whether one uses a wave or a particle language for descriptions—so long as one is aware that both are jointly valid.[1] Several unfamiliar conclusions follow, which it is a merit of the Copenhagen school boldly to have adopted.

Microphenomena are conspiracies of wave and particle properties. But, one must maintain a symmetry between these modes of description, since there is no observational reason for stressing one at the expense of the other. Thus in a two-electron interaction, the description may run: electron creates field; field acts on another electron. But we can always construct a parallel particulate description: electron emits photon; photon is absorbed by another electron.[2] Consider also proton-neutron interaction. In wave notation: neutron creates field; field acts on proton. But we will often say: neutron emits pion; pion is absorbed by proton.[3]

This resolution not to sacrifice either notation itself generates a qualitative appreciation of the Uncertainty Relations. Suppose the microphenomenon—e.g. an electron orbiting—is provisionally described as a cluster of the interference maxima of an otherwise undefined wave group. Then, precisely to locate 'it' at the point-intersection of four co-ordinates would require an infinitude of further waves (of infinitely varying amplitudes and frequencies), so as to increase destructive interference along the line of propagation and 'squeeze' the packet to a 'vertical' line (in the mathematically abstract configuration space, of course). This renders unknowable the particle's energy, which is intimately associated with the amplitude and frequency of the component phase waves. But if we would determine the particle's energy, then the phase waves must be decreased in number, allowing the 'wavicle' to spread 'monochromatically' through the whole configuration space. Thus

$$\psi(x, 0) = (\tfrac{1}{2}\pi\hbar)^{\frac{3}{2}}\iiint_p \alpha(p)(e^{(i/\hbar)p.x})dp.$$

That is, the more narrow $\psi(x, 0)$ is chosen, the broader the bracket of linear momenta p—the quicker the component waves get out of phase, and the 'peaked' packet disintegrates. So also, the square of $|\psi(p)|$ represents the probability of finding our particles with certain momenta if we carry out an experiment measuring linear momenta. This interpretation shows qualitatively the impossibility of determining with precision at t all the 2^n canonical coordinates of a quantum-mechanical system.

Thus when micronature forced a wedding between the concepts classical physics had sundered, three disconcerting consequences emerged as issue of the marriage: (1) Physicists were obliged not to overstress either phase of the new joint notation unless nature dictated doing so (present experience provides little basis for expecting such a dictation. But still, compare note 2, p. 73). (2) They became aware of a profound conceptual limitation—the Uncertainty Relations. (3) They saw the need of a single formalism which could integrate those inharmonious ideas, 'wave' and 'particle', into one powerful algorithm.

Of the 'old' quantum theories of 1913[1] and 1916[2] I shall remark only that they were classical models of the 'Saturnian' atom, proposed by Nagaoka[3] and Rutherford,[4] into which was forced (without reasons) the idea of quantizing electronic orbits. Then De Broglie simply hammered together the wave and particle notations:[5] he apparently had no clear notion of a physical interpretation of these waves; it is often said that perhaps he still has not. Schrödinger took De Broglie's *ondes de phase* literally as classical fields of the Maxwell type.[6] This interpretation was punctured by Born[7] as we shall see. So the elegant wave mechanics of Schrödinger, and the observationally equivalent matrix mechanics of Heisenberg,[8] had to float for a time in a cloud of uncertainty concerning just what experimental sense there is in several parameters connected with $|\psi(q)|^2$. Born dispelled this cloud with the ingenious suggestion that the waves be taken as a measure of the probability of locating particles within a given volume element.[9] Because it was operationally clear, and corroborated by every known experiment, Born's view was quickly adopted, and generalized for multiparticle distributions by Bohr, Heisenberg, Gordon, Jordan, Klein, Pauli and, most significantly, by Dirac.[10]

In 1928 there appeared a great contribution to physical theory, perhaps the greatest of our time. Just as Newton had forged together the independent laws of Kepler and Galileo, all of hydrodynamics and every known fact of astronomy, ballistics and optics, so also did Dirac's theory of the electron blend into one formally beautiful and experimentally powerful theory every idea of the particle physics of the 1920's. He provided a comprehensive and workable model for the hydrogen atom, explained the Compton scattering of electrons, the Zeeman effect, and the empirically required electron spin; all these were blended into an algorithm whose purpose was to achieve a relativistically invariant theory for fast electrons. Dirac's mastery is clear from his elegant adaptation of Jordan's operator calculus (itself a remarkable generalization of Heisenberg's matrix mechanics), so as to make the qualitative uncertainty relations into a formal property of the notation. Dirac took an idea of Graves,[1] as developed implicitly in Heaviside's 'operational' calculus,[2] and already used in quantum theory by Born and Jordan—wherein an ordinary algebra is modified by the law $PQ - QP = n$ (some number other than o). The properties of such a non-commutative system were well understood by 1900. But to translate this formal innovation into a systematic expression of the uncertainty relations implicit in the wave-particle fusion was pure genius. Dirac's paper established quantum mechanics as a unified description of nature. The theory's stature was even more elevated when one of its consequences—first thought a blemish by Dirac (and even earlier (1926) by Gordon[3])—entailed unobserved entities with queer properties. This 'blemish', which Dirac,[4] Schrödinger, Weyl and Oppenheimer[5] tried to eradicate, was seen by Blackett to describe the new anti-electrons which he and Occhialini[6] and Anderson[7] observed in 1932. Dirac's theory did everything; it integrated all available facts, provided a well formed formalism, and was fertile in predictions: e.g. the antiproton and the antineutron have only recently been detected.[8] (See ch. IX.)

Many early objections to, and some present dissatisfactions with, the Copenhagen interpretation arise in part from a failure to appreciate the historical and conceptual role played by Dirac's paper. Here is the notational key to all subsequent quantum physics: yet in that paper (Dirac tells me) the Copenhagen interpretation figured essentially—not as some philosophical afterthought appended to Dirac's algebra, but as apparently basic to every operation within the notation. Feyerabend[9] provocatively suggests that this need not have been so, that it would be possible to have a 'minimum' (i.e. non-Copenhagen) interpretation of quantum theory, and hence of Dirac's paper. We face this issue squarely in chapter VII. But in fact this is not the way in which

Dirac's fundamental paper was actually written: it would not *be* the same paper were its assumptions 'purified' *à la* Feyerabend. It was largely because of its tough-minded and practical suppositions that Dirac's theory had almost complete success, at least before the era of the meson. What critics of the Copenhagen interpretation often fail to see is that merely to insist on an alternative account of micronature without actually providing an idea of one which could work, is to re-invite the chaos which it is Dirac's triumph to have ended. The way to command a practising physicist's attention with counterproposals is to provide a better scientific theory, not simply a restatement of the orthodox formalism plus some metaphysical asides. Perhaps it is possible, as Feyerabend ingeniously moves, to have a minimum statement of quantum theory, with no more 'interpretation' than is required barely to describe the facts. But, rightly or wrongly, this is what Dirac felt he had, and what the Copenhagen position feels it now has, and why it views most counter-proposals as observationally irrelevant superstructures. In 1952 Bohm conceded that his re-interpretation affected no known facts, but only added extra philosophical notions of heuristic value.[1] *Bohr et Heisenberg n'ont pas besoin de cette hypothèse.*

Should philosophers discontinue attempts to develop proposals which counter the Copenhagen interpretation? Not at all! Of course, they might well be somewhat less enthusiastic in their own evaluations of such activity. But what is objectionable is the current practice of introducing 're-interpretations' by way of referring to the Copenhagen school as if it held the field by some kind of authoritative dictatorship; as if several clearly formulable alternative interpretations of quantum theory were being forcibly suppressed in favour of the naïve metaphysics of Bohr and Heisenberg. We should remind ourselves that there is as yet *no* working alternative to the Copenhagen interpretation. It therefore seems a questionable procedure to present every new and tentative speculation as if it were a clear alternative which could easily revolutionize the foundations of physics, if only elder statesmen like Bohr and Heisenberg would stop backing their favourite horse so uncritically.

One might conclude: until you formulate a new interpretation which works in every particular as well as does the old one, call your efforts by their proper name, 'speculations': this makes them no less worth while. Should it be riposted that the Copenhagen interpretation is itself a speculation, then at any event we must distinguish those speculations which have proven themselves to work, in theory and practice, from those which have not yet been put to any kind of rigorous trial.

Recall the analogous uneasiness felt by nineteenth-century astronomers, as discussed in chapter II. Although many thinkers were then distraught by Leverrier's failure to explain the unexpected secular advance of Mercury's perihelion, no one seriously proposed that Newtonian astronomy be abandoned. To have done so at that time would have been to stop thinking about celestial phenomena altogether. One had to provide some equally useful astronomy, or else provisionally accept the otherwise successful orthodox theory.

Quantum theory presents a similar case (although its failings are perhaps nowhere so grave as was the Newtonian impasse). It is arguable that the Copenhagen limitations, far from being the result of philosophical *naïveté*, are built into the very wave-particle duality micronature has forced on us—and built also into the symmetry of the alternative explanations generable in terms of that duality. At least it remains to be shown that this is definitely not so.

Dr Feyerabend would not agree. He distinguishes 'Born's interpretation' which gives the formalism a physical meaning, from 'Bohr's interpretation', which he views as being itself a metaphysical addition to this bare physical theory. Let us suppose that Feyerabend is correct: it would not by any means follow that 'Admitting this implies that we are... free to invent and to consider other "metaphysical" interpretations';[1] for this conclusion would obscure the historical, conceptual, and operationally successful role of the Bohr view as opposed to other accounts. The 'metaphysics' in Newton's *Principia* is not to be rated on a par with the woolly harangues of Hooke and the *unintelligibilia* of Benton. The licence which permits inventing alternative interpretations, a licence which Feyerabend imagines to be granted with the discovery of a metaphysical strain in Bohr's view, is inoperative and unwarranted in the absence of any alternative formalism (and *concrete* experimental suggestions) upon which to build the 'new metaphysics'.

There is a further, experimentally related point: familiarity with the design of apparatus within laboratory microphysics reveals that the only way to learn about fundamental particles is to *interact* with them at our macrophysical level. This is not merely a comment on experimental technique. The proposition, 'To learn anything about a particle we must interact with it', has a logical force comparable with 'No object can move faster than light', or 'There cannot be a *perpetuum mobile* (first type)', or 'A super-Carnot engine is nonconstructible', or 'A temperature-registration of less than $-273°$ C is impossible'. None of these claims state *mere* matters of fact. Each essentially involves the conceptual principles of entire physical theories. Similarly, the proposition: 'One must interact with microparticles to learn about them'. The negation of this, although not logically self-contradictory as with the negation of a tautology, is none the less physically unintelligible, as with all the claims above.

This semantical fact entails precisely what many philosophers of science persist in being unhappy about—that in microparticle physics the data can never come to us packed with invariant properties, 'undistorted' by the observing instrument. Even this way of putting it is somewhat misleading, since it suggests that we could in principle have knowledge of an undistorted micro-observation. *But we do not even know what such an observation would be like.* Hence we are here contrasting the observations we actually make, i.e. the 'distorted' ones, with nothing whatever. So we are concerned simply with microphysical observations: the qualifier 'distorted' adds nothing since it delimits nothing. Data in microphysics can never be less than a compound of the micro-event and some macrophysical system, such as a detector—or just simply ourselves. We entirely lack the concept of a genuine working alternative to this, as a cemetery of dead *gedankenexperimente* proves;[1] for such an alternative would require using a detector whose quantum of perturbation is, and must be, h, to get information about micronature in units smaller than h! This constitutes a conceptual impasse, not merely an experimental one. Interaction is the basic information concept in quantum physics. If the basic unit of interaction is h (which no 'hidden-variable theorist' has yet been emboldened enough to deny), then all information patterns with which we choose to describe the world must also be quantized in those very same units of h imposed by the detector—just as, analogously, we cannot photograph anything smaller than the grains on our photographic film. Anything 'beyond' this is, within microphysics, undetectable—unknowable in principle. Any 'alternative' to this cannot be made intelligible. Yet most early critics of quantum theory, and some contemporary ones as well, seem readily to suppose that they have clear intuitive ideas of what electrons, and protons, are 'really' like, technical and theoretical limitations notwithstanding. They appealed, and appeal, to classical statistical mechanics, where experimental limitations could indeed affect the confidence with which we describe, e.g. each new thermodynamical event. But such practical limitations never alter our concepts of the matter involved; a hidden symmetry was and is always felt to reside behind the statistics. (This confidence resulted from confounding 'phenomenological thermodynamics' with 'deductive thermodynamics'—the latter being entirely developed from the two principles of impotence.) It is, however, *the* feature of experimental microphysics that the degree and manner of the perturbation of the system by the detector is in principle incalculable. This is where the photographic film analogy, and the Heisenberg–Bohr supermicroscope analogy, both break down: for in their cases the apparent technical limitations are eliminable, while this is not so in microphysics.

This is not really a novel situation. We should take very seriously for a moment this reference to statistical mechanics: this is itself something philosophers of physics too rarely do. Boltzmann and Gibbs spring to mind. In his irresolvably statistical gas theory[1] Boltzmann often objects to the Newtonian idea of an observation, as contrasted with what earlier physicists were really entitled to count as such (cf. p. 28). Boltzmann did not feel that gas theory required statistical formulation only because of experimental limitations. Rather, for methodological reasons he viewed statistical mechanics as the primary discipline. It is directly connected with observable parameters. Boltzmann construed punctiform masses and ideal particle populations not as physics' starting-point, but as mere abstractions of only marginal heuristic value. They seemed to him to be but dubious variables hidden within the total laboratory exploration of a physical event. They do, of course, constitute notationally simpler notions than are found in any statistical reformulation. These notions may even be indispensable in calcula-tion, although it is doubtful that Boltzmann would have admitted as much. The simplicity, however, is merely formal, and it should not in any way be confounded with thinking about actual observa-tional data. The development of kinetic theory might indeed have proceeded more smoothly had early mechanics attended more strictly to what is observable—and not exclusively to what it is easier to conceive and to calculate. There are no unique mechanical problems about gas-molecules other than that they must be described as observed *en masse*: they will not submit to being experimentally or theoretically reduced to metaphysically prior, or methodologically heuristic, abstractions. Boltzmann construed every observation as a joint-function of both the design of the apparatus and of the observer's knowledge of the previous state of the observed system.

An even more forceful example, regularly invoked by Bohr, is found in Gibbs' thermodynamics itself.[1] Imagine a hot metal emitting electrons: a measuring apparatus registers, by the blackening of a point of a photographic plate, the emission of any electron with a velocity greater than v. Now delicately adjust the apparatus so that electrons are only emitted very infrequently: this type of measurement of temperature leads to what is often termed an 'objective' determination of at least one of the metal's properties. We express this mathematically by regarding the 'metal system' as a sample arbitrarily selected from a canonical ensemble.

But now, what exactly does 'objective' mean here? It means that any thermometer whatever can be used as the measuring instru-ment—that the measurements do not depend for their values on either the uniquely special properties of a particular thermometer or on the peculiarities of some given observer. Now if the total system, i.e. metal-plus-apparatus, were completely isolated from the rest of the universe (assuming this were possible), it would have a constant energy (whose value would never be exactly known because of the canonical distribution). If, however, the metal is not isolated, as must in fact be the case, its energy varies with time and oscillates in temperature equilibrium about the mean value indicated by the distribution. If, now, the canonical ensemble of the total system follows a Newtonian-type pattern of development, the ensemble evolves, containing as it evolves an increasing proportion of stages in which points on the photographic plate are blackened. The probability that the measuring apparatus will respond at a certain place at t can be exactly calculated, but the exact instant and place of any actual event cannot even in principle be predicted. Were every detail of this experiment known at the onset, in the way LaPlace supposed,[1] we should then be able to determine beforehand each point-blackening, provided only that the experimental system were isolated from the rest of the universe. But in that case any statement of the temperature would be operationally unintelligible. If, however, the experimental system is really connected with, and part of, the external world as physics finds it, then even an ideally complete knowledge of its details would not allow precise predictions, since we cannot possibly know every detail of the remaining universe.

This total experimental system also contains what might guardedly be called a 'subjective' element, despite the counter-assumptions of those who suppose thermodynamics to carry no trace of the observing instrument. In the total absence of a detecting apparatus, the mathematics of the system changes continuously—as outlined above. But now if we introduce a detector, it will suddenly register that a point on the plate is blackened. The formal representation has thus been altered discontinuously, because we have moved from the former situation containing a great range of mathematical possibilities, to a new ensemble containing only the plate blackened at one point. For that event, considered *ex post facto*, the formal statistical representation originally appropriate to the ensemble has no further descriptive utility whatever. This sudden, discontinuous change is not contained, or even hinted at, in the Gibbs' equations characterizing the generalized ensemble: it corresponds closely to the 'reduction of wave packets' in quantum theory. Thus, characterizing an experimental system as a Gibbs' ensemble not only specifies its properties—the characterization itself is also essentially contingent on a detector's presence within that system. Hence the word 'objective' in this context is philosophically somewhat questionable (though perhaps the same may be said of 'subjective', too. For these terms were born in a world quantum physics never made).

So, one can assign an 'objective' temperature to a body only on the basis of evidence concerning the average velocity of its constituent particles, some of which escape from the object (altering thereby 'the body') and are recorded by a detector whose own physical properties are involved essentially in the 'reading'. If one had a LaPlacean knowledge of all the particles involved, that is, a complete tabulation of all the component micro-events and their interrelations, then one could predict the time and nature of every actual recording by the detector. But then, in such a case, one could no longer assign an objective temperature to the body; because the very concepts of temperature and entropy presuppose statistical disorder in the phenomena. 'Objective temperature' and 'actual recordings' are thus mutually exclusive notions, though complementary. The former requires complete randomness; the latter, by determining and defining an actual event, eliminates randomness to that extent. The former is fundamentally indeterminate. The latter is *de facto* determined. Again, every such actual recording, by changing the ensemble, reduces the probability function in the Gibbs' representation—a reduction nowhere written into the equations of motion for the particles.

Against all this classical background, consider quantum mechanics itself. A microsystem can be represented by a wave function, or by a statistical mixture of such functions (i.e. by a density matrix): this corresponds to a Gibbs' ensemble. If, as one expects, the system interacts with the world of which it is a part, only the mixture (or matrix) representation is possible. We simply cannot know all details of the total macrophysical system (i.e. the universe). But if the microsystem is ideally closed, we may have what is sometimes called a 'pure case'—one represented by a vector in Hilbert space. Such a representation is completely 'objective' and the microproperties represented are completely 'objectifiable'. This representation is unconnected with any observer's special knowledge or with any detector's unique reaction. Calculations concerning this 'Hilbert' vector are valid or invalid according to completely objective, standardized rules. But such a representation is completely abstract; it is formal to the point of being physically insignificant. Its mathematical expressions, analogues of $|\psi(q)|^2$, $|\psi(p)|^2$, do not refer to any experimental spaces or to any observable properties: they contain no physics whatever. The microsystem *qua* vector-group-in-Hilbert-space becomes a description of nature only when linked with the possible results of possible experiments. At this juncture, however, we *must* consider the interaction of the system with the measuring apparatus, the detecting instrument, i.e. the observer. We must use a statistical mixture in representing this joint, and discontinuously fluctuating, system.

Nor could this (to some) philosophically painful consequence be avoided by isolating completely the microsystem and the observing apparatus from the macrophysical universe. Its location within, and its connexion with, the universe is a necessary condition for any measuring apparatus to perform its intended function. Its behaviour must be registered within actual experiments if we are to get information from the apparatus. Again, this is not a logical necessity: its denial does not reduce to a statement of the form *P-and-not-P*. If it did, we would not be discussing physics at all, but only the formal properties of an algorithm. Still, such a denial would describe nothing intelligible. It is at this moment inconceivable—i.e. 'systematically unintelligible'—that we should ever encounter a *perpetuum mobile* (first type), or accelerate particles faster than *c*. So also is it inconceivable that in microphysical experimentation the measuring apparatus should both perform its intended function, that is, providing information to us macrophysical beings, and also be isolated from the macrophysical world. The joint system micro-event-plus-detector is therefore describable only by a statistical mixture of wave functions—a density matrix. It inevitably contains statements about the detector or, what is the same to the physicist, the observer.

To deny this would be to rob us of the very concepts needed to experiment in microphysics at all. When detectors react characteristically within an experiment, the original mathematical representation is, as we have said, altered discontinuously. One amongst an enormous variety of statistical possibilities has proven actually to obtain. This discontinuous 'reduction of wave packets', underivable from any form of Schrödinger's equation, is, as in Gibbs' theory, an effect of shifting from mathematically descriptive possibilities to experimental actualities. That is, an actual observation reduces the original ψ, representing a generalized ensemble of possibilities about future particle behaviour, to a new ψ, whose future possibilities will have been altered uniquely and irreversibly by the first observation. This shift from original ψ to a new ψ is not even implicit in the wave equation. One could imagine all this projected back in time, but never so far back that the joint system can be thought of as being separated from its world. Such a separation would be incompatible with the validity of quantum mechanics for closed systems, and hence would leave us no concepts with which we could genuinely think about quantum phenomena at all.

Hence a system considered as isolated from the macrophysical world just cannot be described in classical terms. We may wish to say that the state of the closed system described by a Hilbert vector is 'objective', although it does not in fact obtain; but if we do, the classical idea of objectivity must be abandoned, and an unrealizable abstraction must be put in its place. The description of a micro-system by its Hilbert vector group complements its description in classical terms, just as a Gibbs' description of a body's micro-scopic state complements a statement of its temperature. The description of a micro-event can be effected by classical concepts to just that degree of approximation-to-quantum theory one needs for predictions in any given context. This, indeed, is one version of the correspondence principle as discussed in the last chapter. And it is perhaps the least offensive version. Quantum theory can also be used, however. So the boundary line between the domains (1) object-observed, and (2) instrument-observing, can be pushed indefinitely far toward the latter so as to make the latter area in-creasingly prominent; just as one is always free to regard the temperature of a body x as but a property of some *thermometer* placed in a specified physical context containing x. Hence in this case the statistical nature of the laws of microphysics is again seen to be unavoidable.

So much for the history and the logic of the Copenhagen inter-pretation; now let us consider some representative types of attack on this interpretation.

B

The opposition has moved against Copenhagen on four fronts. The earliest discomfort was almost exclusively concerned with the statistical nature of quantum laws. How should one interpret the ψ function in the fundamental Schrödinger equation?[1] De Broglie's *ondes de phase* were never physically defined.[2] Are they but algebraic fictions, or are they connected with actual physical existants? De Broglie's theory of the pilot wave, a wave supposed to 'accompany' every microparticle, suggests the latter view. But no experiment has ever revealed such pilot waves in electrons, protons, or neutrons. Exit De Broglie.

A purely abstract, mathematical configuration space is not for Schrödinger either. He imagines an infinite number of interfering Maxwell-type waves, whose resultant wave maximum just is the particle in question.[1] For him an electron is an 'energy smear'; $|\psi(q)|^2$ gives a measure of the spread and the intensity of that smear. This view deliquesces in any multibody collision problem. Here Schrödinger must postulate *actual* physical spaces of an indefinite number of dimensions (for N particles ψ is a function of $3N$ co-ordinates)—an idea with no experimental interpretation whatever, save in the 'degenerate' case of one particle, where configuration space and physical space coincide. De Broglie, Schrödinger, and also Einstein and Jeffreys,[2] have always contended that quantum theory has not yet settled down to its normal stride. It is like mechanics was in 1600—phenomenological, unstructured and chaotic. Schrödinger even now still seeks to rewrite microphysics as a kind of classical field theory; others pursue hydrodynamic models in which the singularities are point-localizations within a continuous substratum. These attempts, however, have all proved disappointing, so far. At present we have one way of construing quantum phenomena, and this is in terms closely parallel to the orthodox Copenhagen formulation. Show the working physicist a clear, detailed, physically intelligible alternative to this Copenhagen view and he will readily try it. But physicists tend to use a formalism only so far as its parameters are observationally testable; hence the ψ function is usually taken to measure either the probability of finding a particle within a given volume element, or the probability that certain areas on a target detector will be more affected by particle impact than will other areas. Or, ψ may be related to determinations of the density of particles within a parallelepiped of the particle beam, and so on. This does not minimize the reality of the wave aspect of microparticles. A pencil of β rays scattered by metal foil will leave target patterns describable only by the distribution implicit in the ψ function of the equations appropriate for β particles in such a situation. We possess almost

THE CONCEPT OF THE POSITRON

direct evidence of the correctness of De Broglie's approximation in the diffraction of particle beams by crystal lattices. The resulting patterns closely obey Bragg's law as well as all the other laws of diffraction (by spatial gratings) involved in X-ray diffraction and, indeed, even with the exact wavelengths as required by the law $\hbar K = p$ (K = wave propagation vector). The 'ultimate' property of individual particles responsible for such regular distributions remains for the Copenhagen theorist what it might have been for Newton himself were he now with us: 'I have not been able to discover the cause of these properties of the distribution; it is enough that a particle will more probably strike in one place rather than in another, and that I have a formula for describing this probability'.[1] Or in Bohr's words: 'The entire formalism [of quantum mechanics] is to be considered as a tool for deriving predictions, of definite or statistical character, as regards information obtainable under experimental conditions described in classical terms and specified by means of parameters entering into the algebraic or differential equations.... These symbols themselves are not susceptible to pictorial interpretation.'[2]

Further resistance to the Danish invasion of microphysics takes the form of uneasiness over the uncertainty relations. But it has often been shown that this discomfort usually consists in failing to comprehend the reasons for our needing such a theory at all, or at least in misunderstanding the conceptual structure of the theory. One could generate quantum theory in its entirety from a suitable statement of these uncertainty relations alone. So they cannot be merely a peripheral blemish. They are the logical heart of quantum theory.[3] How else could one understand the signal achievements of Dirac's non-commutative operators?

A third source of discomfort in turn concerns precisely this non-commutative algebra. Some scientists can feel no confidence in a theory whose mathematics are managed according to $QP - PQ = n$, where $n \neq 0$. This is reminiscent of the dissatisfaction many apparently felt with Heaviside's decision to represent dx/dt by p—where p is treated as any ordinary algebraic quantity: which also leads to a non-commutativity.[4] Critics at that time thought the operational calculus to be but a sloppy approximation to some more refined-but-as-yet-uninvented theory. Still, the calculus did its job well, e.g. in alternating current circuit theory, if only

86

one learned the rules of the algorithm. Similarly, with the Dirac calculus: it was more powerful in prediction and in explanation of microphenomena than had been any previous theory. Noncommutativity *per se* is no blemish; here, indeed, it constitutes an ingenious expression of what the data actually oblige us to think. Landé, indeed, may even have an argument which relates noncommutativity to the 'natural' postulates of theoretical continuity and symmetry, as these arise in thermodynamics—especially the second law.[1]

There are genuine formal improprieties within quantum theory, e.g. 'renormalization' and the uninterpretable negative probabilities ('ghost states'). True, some physicists (e.g. Heisenberg) have tried to relate these mathematical inelegancies to the very structure of what is basically the Dirac theory. But even Heisenberg's most recent work remains couched in a non-commutative algebraic framework.

A final kind of discomfort with the Copenhagen interpretation concerns what may be an asymmetry in microphysical explanation and prediction. In classical physics, as we saw in chapter II, explaining X is often symmetrical with predicting X. If I explain X via the laws of a system S by reference to initial conditions α, β, γ, ..., then I might as easily have predicted X on the basis of α, β, γ, ..., and operating on these via the rules of S. Predicting X is thus 'explaining it' before it happens. And explaining X is thus 'predicting it' *after* it has happened. For those who take this to be the paradigm of explanation and prediction, disappointment with the quantum-theoretic situation is certain and inevitable. Thus the great Leibniz would have felt disappointment. He wrote: 'The mark of perfect knowledge is that nothing appears in the thing under consideration which cannot be accounted for, and that nothing is encountered whose occurrence cannot be predicted in advance.'[2]

After a microphysical event X has occurred within our purview, a complete explanation of its occurrence can be given within the total quantum theory. But it is in principle impossible to predict in advance those features of X so easily explained *ex post facto*. This meshes with earlier points (cf. pp. 28 ff.). Expressions of discomfort at this juncture are often merely covert ways of announcing that one likes one's physics to be deterministic, orthodox; in short,

Newtonian. But then would such a person also assent to the second law of thermodynamics? If he does (and what else can he do?) is this not inconsistent, as Landé has argued, with any classical determinism?

C

Heisenberg has examined several typical counter-interpretations of quantum theory. These fall, roughly, into four classes:

(1) The amorphous sighs of yesterday's very great physicists, not one of whom has yet offered a scrap of algebra to back up his elder-statesmanly advice. The indeterministic, statistical character of quantum-mechanical laws fills these worthies with dread for the future of science. This is reminiscent of Hooke's charge, cited in an earlier chapter, that Newton had merely provided a formula for gravity, but had said nothing about its causes—the only subject of scientific importance to Hooke; to which Newton retorted that if Hooke had some mathematical contribution to make, he would listen—but would not hearken to metaphysical poetry alone.[1] (2) Hidden variable or hidden parameter formalisms, each one of which destroys the very symmetry which we have seen to constitute the power, and the glory, of quantum theory.[2] (3) Prose passages which fail to grasp the experimental necessity of interaction.[3] (4) Pure mathematics, offering no hint even of the nature of an experiment which could abrogate, for example, the Uncertainty Relations.[4]

The symmetry sacrifice is deeply significant; and these counter-proposals all sacrifice symmetry in some form. But perhaps a Copenhagen-type interpretation is ultimately unavoidable if things like wave-particle duality and Lorentz invariance are genuine features of nature. Every known experiment tends to support this idea.[5] The theoretician will always favour the theory which saves such symmetries (Yang and Lee included). Schrödinger attempted to restate the Dirac electron theory, eliminating entirely the negative energy solutions. Both Weyl and Oppenheimer showed his attempt to vary with the choice of Lorentz frames; exit Schrödinger (cf. ch. IX). If we decide thus the fate of genuine theories, why not adopt the same criterion for interpretations of theories?

There is a loose analogy between this controversy and that over the nature of material implication in mathematical logic. The 'paradoxical' features of material implication are unavoidable once its truth conditions are identified with those of $\sim p \vee q$. This identification is often termed the cause of the supposed error. It is easy, using well-known deductive rules, to show that $p \supset q$ and $\sim p \vee q$ are mutually deducible. The 'paradox' is thus implicit in these rules. In fact, practising logicians, and students also, never think the rules *per se* paradoxical. Anyone objecting to material implication must therefore consider the rules concerning alternations, the equivalence of alternations and conjunctions, double negatives, conditional and *reductio ad absurdum* proof; and must then say which of these he would abandon or modify and, if the latter, what modification he proposes. Similarly with quantum theory. The 'paradoxes' of the Copenhagen interpretation follow directly upon a cluster of experimental facts, and established formal techniques for dealing with these facts. An objector to this interpretation must also say which facts or operations on the facts he would abandon or modify, and, if the latter, what that modification should be. And here we need detailed, alogorithmic discussion, not just speculations about what might or might not become possible if only certain experimental dreams could come true.

D

Professor Mehlberg has considered[1] Jordan's powerful argument against holding any of the early versions of the correspondence principle.[2] He rejects Jordan's reasoning by first accepting perfect correspondence between classical and quantum mechanics, and then concluding that therefore the latter cannot ultimately be limited in the 'non-classical' ways Bohr and Heisenberg have stressed. This is by now a familiar type of anti-Copenhagen argument which has been used with force, e.g. by Jeffreys.[3] Thus Mehlberg says:

...since the validity of quantum theory is...admitted to range over the whole physical universe, unobjectifiability would be a common feature of all physical concepts should unobjectifiability be an inescapable consequence of quantum theory....

...quantum theory also includes large scale bodies, physical concepts would become unobjectifiable throughout the physical universe and the epistemological consequences of this pervasive unobjectifiability would appear to be crippling for the whole empirical method of acquiring knowledge of physical objects on the basis of observation. In other words, Mehlberg has adopted and exploited the sentiments explicitly expressed by Weyl, and already discussed in chapter IV:

Thus we see a new quantum physics emerge of which the old classical laws are a limiting case in the same sense that Einstein's relativistic mechanics passes into Newton's...when c the velocity of light tends to ∞.

Our arguments in chapter IV purported to suggest that no one is logically obliged to accept Weyl's blunt statement concerning the 'correspondence' of classical and quantum physics; and hence no one need have the perplexities enunciated by Mehlberg above. Only an uncritical acceptance of the Correspondence Principle can incline Mehlberg to suppose that '...since the validity of quantum theory is...admitted to range over the whole physical universe...', therefore all the specific conceptual discomforts of quantum theory must now be taken to affect our understanding of older, more familiar, disciplines. Since this is not true, it cannot count as a hit against the Copenhagen interpretation. As to Mehlberg's position concerning the necessity of 'objectifying' all the properties of fundamental particles, it will be clear to the reader who has come this far that a counter-position is being built up in preparation for our assault on the discovery of the positron.

E

Finally, we must clearly distinguish Philosophical Problems of Quantum Mechanics from Problems of Philosophers concerning Quantum Mechanics. It is a sociological fact that working quantum physicists do not in general bother with problems concerning, for example, the interpretation of the ψ function—not to mention matters concerning 'realistic' or 'non-realistic' interpretations of quantum theory as a whole. This is not because the working physicist refuses to have, or is intellectually incapable of having, philosophical problems; he has plenty. One cannot be exposed to discussions in quantum field theory, and meson theory, without feeling the logical, analytical character of the attendant perplexities. These are not things imported into physics from other academic contexts, such as Faculties of Philosophy. I recall a discussion between Dirac, Heisenberg, and Bethe about whether an otherwise successful algorithm containing inconsistencies (as does the technique of renormalization)—whether this requires immediate examination —or whether danger looms only when the predictions of such a theory fail. Historically this is a conceptually significant problem: remember the ether and cf. pp. 36 ff. Another such problem concerns the interpretation of the so-called 'negative probabilities'.

Heisenberg argues that their existence in a renormalized calculation indicates a flaw in the very technique and approach to quantum theory set out in the Dirac notation. (Renormalization requires non-Hermitean operators which ruin the unitary character of the scattering matrices, require negative probabilities, and invoke physically unintelligible 'ghost' states.) Others—e.g. Bethe, Peierls, Hamilton and Salam—appraise the matter wholly differently.

Another physical practice of philosophical interest is Gell-Mann's almost taxonomical attack on meson theory. Much of the Dirac theory is unsatisfactory in this region: so Gell-Mann has proceeded like Linneaus, or Mendeleeff, or indeed, even like a natural historian—drawing up 'phenomenological' charts of particle properties and allowing generalizations to stand forth from the data. Methodologists' morals about physical theory cannot ignore this 'John Stuart Mill'-type approach. Despite Toulmin's campaign, it does exist in physics and it can be important.[1] Many such difficulties are the daily fare of practising quantum physics. I only point to their existence to mark how little attention most philosophers pay them. This is in itself no reason for philosophers to cease talking about what they wish. But no one should think that because most quantum physicists are unperturbed by the type of question brought to prominence by Bohm, and Feyerabend and Vigier, that they are therefore unreflective, resigned, Berkleyan, computer-ridden predicting machines. What the practising physicist is likely to find difficult in many philosophical papers concerning the foundations of quantum theory, is a facile use of terms like 'reality', 'objectifiability' and 'subjectivism'. One might actually be inclined to suppose that philosophers had settled what a discussion of reality, objectifiability and subjectivism was a discussion *about*. Which reminds me of the question with which an Oxford undergraduate once staggered his tutor: 'What is the external world *external to*?' When I see the full sense, or nonsense, of that question, I may see also how quantum theory is going to help, or be helped by, the cracking of such old philosophical chestnuts.

Historically, if not logically, the Positron discovery and the Copenhagen interpretation are intimately connected. In this, our fifth step towards the idea of the positive electron, we have tried to indicate the nature of this connexion. Since quantum theory is generated from the conjunction of two apparently incompatible classical theories, those of wave propagation and of particle interaction, one would expect it to have an unusual interpretation. Undeniably, the Copenhagen interpretation is unusual. So is the positron discovery with its attendant theory of anti-matter. In a sense, the unorthodox features of both were mutually reinforcing; to the degree to which the new quantum theory required a wholly non-classical interpretation, and to the degree to which Dirac's work was written against just this non-classical background, to that same degree physicists were prepared to entertain hypotheses of wholly non-classical particles—like the positron. But more than this, the independent-yet-interdependent approaches to the positron discovery (e.g. of Anderson, Dirac and Blackett) are themselves a reflexion of the independent-yet-interdependent features of quantum physics itself (e.g. the wave-particle duality). It has been the objective of the Copenhagen Interpretation to present this interdependent independence in a plausible and intelligible manner.

CHAPTER VI

SOME CAUTIONS

We have seen that within the past decade there has grown up an acute and articulate group of critics of the 'Copenhagen Interpretation'. The writings of Bohm, Bopp, Janossy and Feyerabend must be taken seriously. The future of important themes within the logic of science may rest with these individuals. But in their own writings they have occasionally matched the inelegancies of Bohr and Heisenberg with their own. The present chapter presents a quintet of considerations which may lead (1) to some re-assessment of the issues between Bohr, Heisenberg and their critics, and (2) some further understanding of the context of the positron discovery.

A

Historical

It was just urged that almost all practising quantum physicists of the 1928–35 period inclined towards some form of the Copenhagen Interpretation. They would have claimed then, as they now do in retrospect, that this interpretation influenced even their manipulations of the formalism. They were orientated in Copenhagen terms and they evaluated proposed lines of research accordingly: 'But, this was unnecessary. There always has been a "minimum interpretation" for Quantum Theory which adds but the barest physical meaning to the formalism—certainly nothing so extravagant as the entire Copenhagen Interpretation *en bloc*.' This rejoinder is much like the things recently written with frequency and force by Bohm, and Feyerabend—the two most articulate critics of the Copenhagen Interpretation of Quantum Theory.[1]

Suppose that (for the sake of argument) this point is conceded to Bohm and Feyerabend. From this it does not follow that Dirac, Bohr, Heisenberg, Pauli, and others did themselves actually use such a minimum interpretation.

Perhaps it *can* reasonably be argued that the Copenhagen Interpretation is not a logically necessary supplement to the orthodox quantum formalism. But such an argument would differ from the argument that the physicists of the Twenties and Thirties ought not to have interpreted quantum theory as in fact they did. There are good historical reasons for their having done so. One cannot allow to go unchallenged the suggestion that, because the Copenhagen Interpretation cannot be shown to be the *only* interpretation for quantum theory—that therefore the reasons Bohr and Heisenberg had for adopting this view were not good reasons. This suggestion would have the absurd consequence of characterizing every interpretation of every theory as resting on poor reasons, since every genuine theory in physics must be open to more than one physical interpretation.

Dr Feyerabend is careful on this point. But there is a tendency in his writings to read the (supposed) existence of alternative interpretations of quantum theory back into the past. The initial adoption of the Copenhagen Interpretation is thus made to seem like a perverse refusal to recognize equally effective alternative interpretations. As a matter of historical fact, this adoption was a triumph in that it introduced an orderly interpretative pattern into an inchoate cluster of experimental facts.

Compare with this Newton's interpretation of the equations within his optical theory. His interpretation likewise lacks any exclusive necessity (which, again, must have been the case in order for that interpretation to have injected into the theory's bare formalism any informative content). But this logical fact does not entail that Newton's theory of Fits of Easy Transmission and Reflexion played no conceptually significant role in developing his optical theory. The opposite is the case, as we saw in ch. 1. Similarly with the Copenhagen Interpretation and the actual historical development of quantum theory.

But perhaps Bohm and Feyerabend need not worry about this merely historical issue, save in so far as it signals a possible misunderstanding of their views on the part of some readers.

B

The Copenhagen Interpretation and the Bohr Interpretation

Feyerabend and Bohm are concerned with the inadequacies of the Bohr Interpretation (which originates in Copenhagen). Both understress a less incautious view, which *I* shall call 'the Copenhagen Interpretation' (which originates in Leipzig, and presides at Göttingen, Munich, Cambridge, Princeton, Berkeley, and almost everywhere else). Both writers are upset by Bohr's claims as to the finality of the descriptive dualism within orthodox quantum theory; they are similarly disturbed by the more formal analyses of von Neumann which terminate in the thesis that, within the algorithm he [von Neumann] has set up as a description of microphenomena, the hypothesis of further 'hidden' parameters, which might resolve the dualism into a comprehensive wave notation or a comprehensive particle notation, is impossible. Thus Bohm's conjectures about subquantum, or subelectronic, field effects would be ruled out in advance by von Neumann's 'proof'. Vigier's more recent work would have been dealt with similarly.

More than this, Feyerabend objects to such ideas as that the macrophysical system (observer, detector, or some other particle configuration), with which the specimen particle is entwined, somehow causes, or forces the specimen particle both to have its observed properties and also to endure transitions from one state to another.

It is important so to point out the difficulties in such 'majority' views and to remark also the many situations within physics which seem intuitively incompatible with Bohr's position.[1] Bohm and Feyerabend are also correct (as is Mehlberg) in claiming that there is no binding reason why an algorithm dealing with pure quantum states need be interpreted as Bohr suggests.

However, there is one condition central both to the Bohr interpretation and what I call 'The Copenhagen Interpretation'. This condition is that all empirically meaningful statements within quantum theory are ultimately about 'phenomena', by which Bohr means: the-specimen-particle-plus-the-detecting-apparatus-without-whose-mutual-interaction-no-particle-could-even-be-apprehended. This corresponds to what Heisenberg calls 'the necessity for statistical mixtures'. These mixtures are the inevitable consequence of treating the solutions for the Schrödinger equations as possible descriptions of what is observed in the laboratory.

Now consider together:

(1) The principle that every microparticle, in so far as it is observable, is but a selected portion of some more complex system, whose description is the quantum theoretic analogue of the Gibbs' ensemble of kinetic theory.

(2) The further (although perhaps equivalent) thesis that all references to transition probabilities between 'pure' states of a microsystem—where these references in no way involve mention of an observer or a detector—concern nothing more than the formal operations permissible within the algorithm; in no sense would such a reference constitute an observationally testable statement of quantum physics (as opposed to quantum mathematics).

(1) and (2) constitute a minimal account of what most physicists understand as the 'core-meaning' of the Copenhagen Interpretation. This account will be described in more detail later. Feyerabend, Bohm, Vigier and Mehlberg would, of course, find this account a much smaller and more elusive target to aim at than are the *ex-cathedra* utterances of the melancholy Dane. Bohr's further verbal excesses, e.g. his talk about the detector *forcing* the specimen-particle to have the properties it exhibits in a particular interaction experiment, these are inessential to the above version (i.e. (1) and (2)) of the Copenhagen Interpretation. This version is stronger than Bohr's because its claims are fewer and hence its vulnerability less; but I realize that this might also establish just the opposite— that (1) and (2) are therefore weaker, because less vulnerable, than Bohr's version.

This slightly 'softened' (or 'stiffened') rendition of the Copenhagen Interpretation is more transparently connected with the duality principle than is Bohr's view. It simply urges this: that whether or not one chooses in one's prose accounts to stress the wave aspect or the particle aspect of an observed phenomenon, the representation of that phenomenon within quantum theory will always have the form of a density matrix (i.e. a statistical mixture of wave functions analogous to the mixture within the corresponding Gibbs' ensemble).

Moreover, this 'adjusted' view meshes with one obvious form of complementarity, viz. that in one's accounts of observed phenomena, while the basic symmetry of quantum theory allows one to stress either the wave or the particle aspect of the observation as one finds convenient, none the less one cannot stress *both* aspects at once. There is a conceptual tension between these two modes of description dating back, at least, to Grimaldi, Huygens and Newton.

This suggests that it is misleading to make it sound as if all opponents of the Bohm–Vigier–Feyerabend type of re-interpretation are 'Copenhagen Theorists' in the incautious Bohr interpretation sense. It is possible both to reject as unnecessary to the structure of quantum theory the science-fiction excesses of some of Bohr's prose, and also to reject proposed re-interpretations of the Bohm–Vigier–Feyerabend type. (Landé's work may in fact steer between Bohr's *Scylla* and Bohm's *Charybdis* in just this way.) Surely Bohm's theories have encountered independent difficulties which are just as offensive as those which Bohr's view encounters. There *is* a plausible interpretation of quantum theory, the essentials of which are set out in Heisenberg's essay in the well known Niels Bohr volume,[1] which is not prone to Bohr's excesses, which is properly called a 'Copenhagen Interpretation', and which also provides good arguments against specific counter-interpretations offered by Bohm, Bopp, Fenyes, etc. To say this, however, is not meant to deter the general *programme* of seeking such a re-interpretation.

C

Elementary quantum theory

Bohm and Feyerabend seek a re-interpretation of something they call 'elementary quantum theory' (EQT).[2] But this seems only an arbitrary and artificial abstraction out of a more comprehensive albeit less elegant theory—usually called just 'quantum theory'. This latter term is usually meant to include quantum field theory (QFT). It is QFT which dominates research in particle physics today, not the portions of von Neumann's 1931 formulation arbitrarily selected by Bohm and Feyerabend as 'EQT'.

But they may talk about what they please. If it is non-relativistic quantum theory's connexions with the Bohr interpretation they are interested in, *à la bonne heure*. Still, caution is necessary. Every exposition of Quantum Electrodynamics, and of QFT generally, begins as an extension and a completion of the programme of EQT sketched by von Neumann. Most physicists view that sketch from the other direction (from QFT back to the EQT) and would regard any division between the two as arbitrary—concerning both *where* the division is to be located, and even *whether* any such division is feasible.

Nor can one speak too readily of EQT 'as formulated by von

Neumann', or even selected portions of that exposition. Because, whatever Bohm and Feyerabend may admit into and exclude from EQT, von Neumann's book treats several problems connected with radiation and electromagnetic effects as *its* responsibility. No theoretical physicist of that period could have refused to discuss such matters. But these problems are not adequately resolved in von Neumann's treatise. Perhaps full resolution would have been impossible at that stage. Von Neumann's field-theoretic perplexities are better handled via the theories of Dyson, Schwinger, Tomonaga and Feynman. So, by the very problems von Neumann raises in his exposition, EQT is but a programmatic sketch of something more comprehensive; it is but an arbitrarily determined subtheory within QFT. Most field theorists suppose that, in principle, everything expressible within EQT must also be expressible in QFT, but not vice versa.

If, however, it is plausible to suggest that QFT comprehends EQT, then the status of the former will surely be relevant to assessing the status of the latter.

The status of QFT is readily determined. As a mathematically well formed theory it does not exist. The formal inadequacies of this algorithm are well known. Just one example: consider the formalism fashioned by Heitler in his book *The Quantum Theory of Radiation* (1936). There are a variety of solutions for the wave equations in Heitler's formalism in which the integrals radically diverge. This renders infinite the number of possible answers to Heitler's precisely formulated physical questions.[1] To get around this difficulty Bethe, Tomonaga and Schwinger ingeniously proposed 'renormalization' which, simply put, consists in the artifice that from this infinitude of solutions the theoretician is free to choose—on mathematically extraneous grounds—a finite selection for further consideration. Not only is this an extraordinary mathematical practice, it has disturbing physical consequences as well: (*a*) it renders the operators of the system non-Hermitean, which (*b*) immediately destroys the unitary character of the *S*-matrices (scattering matrices); hence (*c*) we may find ourselves calculating the probability of a transition occurring within a volume element even where it cannot be supposed that within that volume element there actually is a particle to change from one state into another. (*d*) This leads to the physically unintelligible notion of *negative* probabilities; one finds oneself talking about 'ghost' states and other fantasies, as we saw on p. 91.

QFT considered as an algorithm cannot stand on its own feet. But if QFT includes EQT then one must press the question: Is EQT even satisfactory as an algorithm usable in physics? If a negative answer to this has any merit, then any further question concerning the *interpretation* of such a weak formalism would depend upon first doing the necessary repair-work to the algorithm. This must precede any attempts finally to decide (*à la* Feyerabend, Bohm *and Bohr*), what EQT is ultimately all about. For suppose that I took an obviously inconsistent algebra, and often found it useful as a calculating aid in physics; would I not be ill-advised to quarrel with anyone about what the algebra-as-thus-used was *really* all about? Such a quarrel ought to follow only *after* tidying up the mathematics. This, in fact, is what most practising theoretical physicists seek to do. Perhaps this explains, although it may not justify, why physicists remain somewhat indifferent to issues of the Bohr versus Bohm–Feyerabend variety. Such issues they regard as ill-founded in the absence of any solid mathematical foundation to serve as the base for discussions of 'interpretations'.

That is, if substantial algorithmic difficulties remain in EQT, which would be the case if QFT can indeed encompass it, then questions of interpretation may be insoluble just because all questions of interpreting a formally inadequate theory must remain insoluble.

When Newtonian mechanics encompassed Galilean and Keplerian mechanics, a consistent interpretation of the latter two became feasible precisely because the former was algebraically adequate. Had this not been so, e.g. had Celestial Dynamics been riddled with mathematical *unintelligibilia*, physicists of the eighteenth century would rightly have postponed the question of interpreting the equations of Galileo and Kepler until Newtonian Dynamics had been hammered into a formally acceptable shape. Where this cannot be done, the interpretation issue cannot be sharply put.

Of course, someone might have decided, even then in the eighteenth century, to narrow the field of inquiry radically. He might have elected to treat Galilean and Keplerian mechanics as a closed system, discussable in isolation from the general dynamical context of the *Principia*. This move at that time would have had the same kind of justification that Bohm and Feyerabend now have for restricting the discussion to EQT, when QFT as a whole remains in Heraclitean flux. In short, Bohm, Vigier, Feyerabend, etc., have cut a 2-inch square picture out of a very large canvas; they have then enlarged the tiny square to the size of the original canvas. This is a large target. But one might reasonably refuse to throw the first dart at it on the grounds that such an enlarging procedure may itself be far from unquestionable.

Bohm and Feyerabend might seek to deny that EQT is contained within a QFT. The grounds might be that there is no formal proof that this is so. But, similarly, there is no formal proof that quantum mechanics contains classical mechanics as a limiting case, although this is very often claimed. None the less, in just the sense in which EQT has forced us to reassess Newtonian physics as a formalism, so also QFT may yet force us to reassess EQT as a formalism.[1]

D

Symmetry

Let us grant *pro tem* that EQT can be treated as a closed, non-relativistic system. After all, philosophers of science may discuss what they choose. There remains the one final property of the orthodox theory which we described in the last chapter, and which is so fundamental to the theory that any decision to ignore or change this property just is the decision to ignore or change the theory as a whole. This is the 'symmetry property' of EQT.

Nature (i.e. the findings of experimental-observational physics), gives us no conclusive reason to decide now in favour of singularities, or in favour of fields, as constituting 'the ultimate reality', if the reader will forgive this medieval talk. Davisson–Germer distributions of β-rays are just as fundamental observationally as is the Thomsonian defection of a β-beam when it enters a transverse magnetic field. These are equally basic observations, and neither will be reduced to the other! Similarly, the general formalism of EQT is not biased either in favour of fields or in favour of singularities—waves or particles. The theory is neutral on this: we remain free to describe a microphenomenon, in prose, so as to stress either its singularities or its field properties, as we see fit, provided we never attempt to stress both equally and at the same time. This is elegantly shown by von Neumann and, in a different way, by Dirac.[1] The celebrated 'von Neumann proof', of the impossibility of any 'hidden' parameters ever augmenting quantum theory, is the argument that if we opt for a single ultimate microphysical reality, either in the form of a singularity or of a field, either as a granular configuration or as a hydrodynamic substratum, then we will by that very option bias EQT and destroy the symmetry which is now its salient feature. Such an option, in short, will not supplement the old theory; it will substitute a completely new theory. An absolute change changes absolutely: that is von Neumann's 'proof'.

Moreover, such a substitution will have been effected without any clue from nature that this should be done. If particle collisions were discovered to be fractures in a more fundamental plenum, or if Laue distributions were revealed only to be a Coulomb repulsion between singularities, then nature would indeed have disclosed to us a new parameter heretofore hidden. But nature has not done this. It has not even done anything like this. One need not join with Bohr and Heisenberg in pronouncing that nature could not do this. But still, such a possible disclosure seems monumentally unlikely.

What then justifies altering the symmetry of EQT when nature has not requested any such change? To answer: 'so that we can reinstate EQT as deterministic via hidden [classical] variables' is to force the argument into a *petitio principii*. If one is prepared to make *this* change without any help from nature just to regain the heuristic comforts of classical determinism, why then one might as well have clung to classical punctiform mechanics twentieth-century microphenomena to the contrary notwithstanding. Classical mechanics is just false as an account of the microworld, of course. But it is comfortable. Hidden parameter theories are comfortable in much the same way. However, since they must be non-symmetrical as between singularity and field accounts of nature, they also must be adjudged false—at least on all the present evidence (which is as much as we can ever say *vis-à-vis* factual falsehood).

The point is this: even though counterfactual in the respect noted, a hidden parameter hypothesis cannot be supposed to add something to von Neumann's original theory which he mistakenly left out.[1] Since the symmetry condition is fundamental to the structure of both the von Neumann and Dirac formulations, a hidden parameter theory (which must contain some form of the negation of this condition) will be not just an addition to EQT, it will be a totally new and logically different theory.

Where is there to be found an example of this new and different theory worked out in the detail required to discuss, describe, explain and predict microphenomena? Nowhere. We have only promises. Plausible and ingenious promises agreed, especially those of Vigier; but nothing yet approaching an alternative, non-symmetrical theory. This being so, it explains why some modified statement of the Copenhagen Interpretation (as against the Bohr Interpretation), may be adhered to not because anyone supposes it necessarily true, and not because it is part of a dictatorially enforced orthodoxy emanating from Copenhagen headquarters—but just because physicists, being no less reasonable than other men, will not abandon imperfect tools in their hands for the optimistic expectations in philosophers' minds. Let us develop this as a fifth 'caution'.

E

The unstressed strength of the Copenhagen Interpretation

The class of incontrovertibly true statements includes these two kinds of assertions: 1, those which are such that their denials are demonstrably self-contradictory; and 2, those which are such that their denials, although consistent, describe nothing intelligible to think about, or with. 'All bachelors are male' is true in sense 1. 'All bachelors weigh less than the sun' is true in sense 2. 'No triangle is quadrilateral' is incontrovertibly true in sense 1. 'No β particles of precisely determined energy can have precisely determined co-ordinates' is incontrovertibly true in sense 2.

Bohr and Heisenberg sometimes speak of the permanence for physics, and of the conceptual novelties embodied in the Copenhagen Interpretation. They find the idea of an EQT 'emancipated' *à la* Schrödinger, De Broglie, Einstein, Jeffreys, Bohm, Bopp, Fenyes, Bunemann, Janossy, Vigier and Feyerabend simply inconceivable, incontrovertibly false. But in what sense 'inconceivable'; in what sense 'incontrovertible'? Surely not in sense 1. Only in careless moments (which ought not to interest philosophers) would Heisenberg claim a necessity for the Copenhagen Interpretation which would make him say that dissenters were just contradicting themselves. Whatever else this Interpretation may be, it is not analytically true.

How then do, for example, Feyerabend's allusions to the history of science impinge on the Copenhagen theorist's claims? Had Heisenberg supposed his interpretation of EQT to be incontrovertibly true in sense 1, it would be salutary to review past physical theories, and their interpretations, as Feyerabend has done. Many of these[1] were once thought to be incontrovertibly true in this sense, only later to be replaced by new theories and new interpretations. Feyerabend, Bohm and Vigier view the Copenhagen Interpretation against this backdrop. They conclude, of course, that Bohr and Heisenberg express themselves too comprehensively and inflexibly, the state of EQT being what it now is.

But this misses the real force of the Copenhagen viewpoint. It may be that the suggested permanence of this conception derives not from any positive conviction that alternative views are demonstrably inconsistent, but rather from the lack of any positive conviction that there is a consistently intelligible alternative.

This sharp point is often made by Heisenberg rhetorically: 'What detailed concepts am I to entertain when invited to dwell on the shortcomings of the Copenhagen Interpretation? Describe precisely some alternative way of thinking about elementary particles which is supposed to be operationally and conceptually just as effective as that interpretation.'

There is as yet no algebraically detailed and experimentally adaptable answer to Heisenberg's challenge. Nothing in the past or present utterances of Bohm, Feyerabend or Vigier can claim here and now to be a working alternative to the Copenhagen Interpretation.

It is not enough just to note that since science evolves, the Copenhagen Interpretation may change with it. For this is to note no more than that physics is a body of empirical knowledge, and not a formally closed algorithm. No physical theory is incontrovertibly true in sense 1. Who would deny this? Not even Eddington, at the zenith of his apriorism.

Parts of our physical thinking, however, may now be incontrovertibly true in sense 2! Again, this is a conceptual point, not a psychological one. That χ is not thinkable is not simply another experimental datum. This fact affects everything philosophers *can* say about χ and its context. To say 'we do not know now what a *perpetuum mobile* would be like' is not like saying 'we do not know now what Venus' surface is like'. Similarly, to say 'we do not know now what a Quantum Physics without duality, complementarity and uncertainty would be like' is not like saying 'we do not know now what a Quantum Physics without divergence difficulties, renormalization inelegancies and negative probabilities would be like'. The latter features are irradicable; the former are not—not without the wholesale scrapping of everything we now recognize as Quantum Mechanics. That these features of Quantum Mechanics differ in type must be appreciated if one is to appraise the Copenhagen argument at its strongest. Heisenberg is not making a remark about physics' historical development ('We changed it all and it will never again differ from what we made it into'). At its best the Copenhagen Interpretation reflects the conceptual possibilities remaining open to practising physicists ('We are now obliged to think about microphenomena in this way; we know of no alternative way of thinking about them, and anticipate none, nor has anyone yet come forward with a workable alternative suggestion').

Thus, recognizing that a physical theory and its interpretation is never final (*qua* logically closed) is not the same as recognizing, in addition, well developed alternatives to that theory, and then asking which alternative is true. Newton, Fresnel, Young, Fizeau and Foucault—Planck, Einstein, Compton and Raman—they were in this position of choosing between well developed alternatives (in some cases they actually developed the alternatives). Elementary particle theorists today are definitely not in this position. Generally, they are not even looking for alternatives. The practical difficulties of microphysics consist not in reinterpretation, but in algebra. Dirac has even remarked that we still need a formally workable theory of the electron. Ask the nearest Synchrotron operator; see if Dirac exaggerates.

In the absence of a detailed alternative way to think about Quantum phenomena, what is one being asked to reflect upon

SOME CAUTIONS

when invited to consider the inadequacies of orthodox Quantum Theory, and the possible advantages of the as-yet-undeveloped counter-views of Bohm, Bopp, Vigier and Feyerabend? Nothing, really. Indeed, the history of physics can as easily be pressed to support this point as to support Feyerabend's contention. Celestial mechanics was not abandoned wholesale when the completely 'counter-Newtonian' precession of Mercury's perihelion was discovered by Leverrier in the 1840's. Indeed, as we saw on p. 27, Leverrier went so far as to invent the improbable planet Vulcan to account, à la Neptune, for the anomaly. He had to settle, in the absence of Vulcan, for interplanetary dust—and ultimately for nothing at all. But so long as there was no alternative theory to think about planetary motion with, the shortcomings of orthodox celestial mechanics were gratefully tolerated. The philosophical visionaries of the 1860's and 1870's, those who speculated about a fundamentally different astronomy, cut no ice with practising astronomers. More than sixty years elapsed before Leverrier's perplexing discovery was even partially explained; and then not until a fully implemented alternative theory was available to do all the work the former theory had done, plus this extra bit as well. Where is the alternative to the Copenhagen Interpretation which can be spoken of thus?

Perhaps the Copenhagen Interpretation is somehow comparable in this respect. It will not be abandoned until it is completely replaced. And it will not be replaced by anything short of a completely detailed, algebraically articulated theory. The prospect of such a renaissance in the immediate future seems faint, Vigier's ingenuity, Bohm's industry and Feyerabend's energy notwithstanding.

To quote the Bohm of 1950:

...no experiment has yet shown the slightest trace of such hidden variables...[and] there are strong theoretical arguments which make it unlikely that such hidden variables exist....[p. 29.]

...the general conceptual framework of the quantum theory cannot be made consistent with the assumption of hidden variables...no completely deterministic mechanism that could explain correctly the observed wave-particle duality of the properties of matter is even conceivable. Before we could justify the assumption of such a completely deterministic underlying theory, we would have to prove first that the quantum theory is not in complete accord with experiments...in no case has it ever been found to contradict experiment....Until we find some real evidence for a breakdown of the general type of quantum description now in use, it seems, therefore, almost certainly of no use to search for hidden variables. Instead, the laws of probability should be regarded as fundamentally rooted in the very structure of matter.... [p. 115.]

Unfortunately, such an experiment [one which would violate the Uncertainty Relations] is still far beyond present techniques, but it is quite possible that it could some day be carried out. Until and unless some such disagreement between quantum theory and experiment is found, however, it seems wisest to assume that quantum theory is substantially correct, because it is a self-consistent theory yielding agreement with such a wide range of experiments not correctly treated by any other known theory.' [p. 623.]

Since writing these sentiments (strong, perhaps, even within the class of Copenhagen theorists) Bohm has, of course, re-oriented his position in the ways which have elicited the writing of this chapter. Still, it is not clear that Feyerabend and the 'new' Bohm have really met the 'older' Bohm's challenge. However.... What then, is the message to carry to Copenhagen? That some new interpretation of microphenomena is ready at hand? Not yet. What then? Only that the Copenhagen Interpretation may not be perfect in every respect. But who denies this? And in the absence of a comprehensive alternative way of looking at microphysical phenomena who will say exactly where imperfect? But then what is the Copenhagen Theorist being asked to worry about, beyond what he has never ceased worrying about, namely the *de facto* utility of his interpretation?

This step in our argument follows on after the one before. Chapter v set out sympathetically the Copenhagen Interpretation of quantum theory. Here we have considered that interpretation in light of some recent attacks. The entire plot of the positron story is set into this context; to fail to feel the historical role of that interpretation would certainly affect one's understanding of the Anderson–Dirac–Blackett revolution in modern physics. Indeed there is an intimate connexion between the conceptual freedom the Copenhagen Interpretation brought into microphysics, and the unorthodox and non-classical notions of anti-matter formed in the 1928–34 period. The sequel will bear this out.

CHAPTER VII

UNCERTAINTY AGAIN

Two contentions within Niels Bohr's early outlook were of central importance in the forming of the positron concept:

(a) Quantum theory will always have to work with dynamical states which are only partly well defined.

(b) These dynamical states must be regarded as irreducible and ineradicable relations between the microsystem and some appropriate measuring device.

Let us elaborate (a).

There exists a veritable cemetery of dead proposals purporting to show how the 'classical state' of a microparticle might be precisely specified, Heisenberg's Indeterminacy Principle notwithstanding. These have been surveyed. Some of these proposals seek experimentally to abrogate the Uncertainty Relations. Others try, by subtle and skilful manipulations of the quantum algorithm, to reinstate talk of 'superstates'[1] as both intelligible and plausible. We will offer here two more inmates for this cemetery. They are *prima facie* quite plausible, and will be presented with an initial air of conviction. However, for a host of reasons connected with the theses of the last three chapters, the following examples cannot 'come to life', and that this is so reflects, as with a mirror's image, the dynamic growth of the anti-matter hypothesis.

Consider what experimentalists call an 'α star'. All α particles out of a specific decay process have the same range. Being ejected from an α emitter under the same circumstances, these particles have identical energies. Hence they are seen as a 'fan', or 'star', of tracks, all moving from a common source.[2] Suppose someone undertook precisely to locate one of these particles. By a simple experimental procedure the particle α_1 is discovered to have coordinates x, y, z, at t. This position can be sharply determined, and represented in approximately punctiform terms. But the energy of α_1 is already exactly known; it is precisely the same as the energy of any other α particle out of the same emitter. And since, by an independent experiment, the energy for any α_n can be precisely determined, it follows that *ex post facto* α_1 could be precisely described as having had, at t, co-ordinates x, y, z and energy e—the precise value of e having been determined for α_1's 'twin brother' α_n, via the Principles of Energy Conservation, and Identity.

Consider now a cognate case, slightly more formally expressed. Let an α particle be emitted, e.g. by the spontaneous decay of an unstable isotope, from position x_1, y_1, z_1 and time t_1. Stop the particle in a photographic plate at x_2, y_2, z_2 and t_2. Errors at the point of origin are $\Delta(x_1, y_1, z_1)$ and Δt_1. One of these two parameters, e.g. $\Delta(x_1, y_1, z_1)$ [hereafter written simply as ΔX_1], is clearly independent; it can be made as small as one chooses. Now a lower limit is fixed for Δt_1 by the uncertainty p (as usually understood): if, now, the velocity of the particle is v, then

$$\Delta v > h/m\Delta X_1$$

and

$$\Delta E = \Delta\left(\tfrac{1}{2}mv^2\right) = mv\Delta v > hv/\Delta X_1.$$

The corresponding Δt_1 is $h/\Delta E = \Delta X_1/v$. Hence, Δt_1 can be taken as equal to $\Delta X_1/v$—as might have been expected even without the Uncertainty Principle. Now, at X_2, t_2 we have errors ΔX_2, Δt_2, $= \Delta X_2/v$. Hence, when the α particle arrives at X_2, its ΔX is ΔX_2. And we can compute its momentum to be

$$p = m(X/t) = m(X_2 - X_1)/t_2 - t_1.$$

The error, Δp, is described as follows:

$$\Delta p = m\left\{\frac{\Delta(X_2 - X_1)}{t_2 - t_1} - \frac{X_2 - X_1}{(t_2 - t_1)^2}\Delta(t_2 - t_1)\right\}$$

$$= m\left\{\frac{\Delta X_1 + \Delta X_2}{t_2 - t_1} + \frac{X_2 - X_1}{(t_2 - t_1)^2}(\Delta t_1 + \Delta t_2)\right\}$$

$$= m/t_2 - t_1\{\Delta X_1 + \Delta X_2 + v(\Delta t_1 + \Delta t_2)\}$$

$$= \frac{2m(\Delta X_1 + \Delta X_2)}{t_2 - t_1}.$$

Clearly, this becomes arbitrarily small by making $t_2 - t_1$ very large. Consequently, after t_2 we can know exactly what was the particle's position (at that time, t_2)—namely X_2. And, from knowing the initial position, X_1, plus the time required for the α particle to reach X_2, we can compute the energy with maximum classical precision. Would not this precisely characterize the α particle's classical 'state' at t_2? Can we not say we *know* the α particle's position and energy, simultaneously and with exactitude?

Both of these examples suffer from the same defect. Both are completely degenerate. We determined the energies *ex post facto*, as if we had run a bicycle into a wall. This information is useless for future predictions concerning single α particles, or single bicycles; that very energy which can be so precisely computed in these contexts is incalculably different after the position-measurements—at least in the case of the α particle, if not the bicycle. For, either the particle is 'killed' in the plate, or—if it passes beyond—its energy must remain completely indeterminate as a result of our learning where, in the plate, it penetrated at t. The α particle thus 'fixed' in the past is not even describable by a wave function (in the past).[1] Indeed, the very language used to describe in prose situations such as these are no proper part of quantum theory at all. In any event, examples like the above no more 'falsify' the quantum-theoretic claim that α particulate states are only partly well defined, than would finding a deuce on a just-cast die contradict the claim that the die is unbiased. Knowing precisely what the die reads now is useless for predicting what the die will turn up next. Knowing exactly the 'state' of an α particle at X_2, t_2 is useless for predicting its state at $X_2 + \Delta X$, $t_2 + \Delta t$. The deuce, and the α particle, are 'dead' at $X_2 t_2$; the analogies between the cast die and the spent particle are very instructive when viewed against the requirements of the probability calculus appropriate to each case. But quantum theory is pre-eminently concerned with 'live' particles having energetic futures: in this it is like a science of perpetually tumbling dice. Every *gedankenexperiment* which seeks more knowledge about a particle than quantum theory is able, or designed, to give, succeeds only (as with the foregoing examples) in proposing a way of killing the particle, or of erasing our future knowledge of it.

An analogous misconception is this: quite often quantum statistical mechanics is compared directly with classical statistical mechanics, the objective being to show that partial definition of states is no more a matter of *theoretical* principle in the former case than in the latter.[2] Of course, the individual elements within the 'ensembles' computed in classical statistical mechanics cannot in practice have their states experimentally determined; that is, we cannot determine the state of this particular molecule in a thermally excited gas. Still, it always *makes sense* to speak of those individual elements, i.e. molecules, as *having* a state which is in principle precisely determinable. It is just this description of the elements within the 'ensembles' of quantum statistical mechanics which is ruled out by the logical structure of the only extant theory. Hence any analogy between these two kinds of statistical mechanics must be very carefully drawn. Since the applications of statistical techniques within quantum theory depend on rejecting an assumption on which classical statistical mechanics is founded, viz. that microstate descriptions are in principle possible, it is unlikely that any gross analogy between the two will ever eliminate this assumption. Indeed, these two species of statistical mechanics are as different conceptually as are classical mechanics and quantum mechanics themselves. So the latter two cannot be bridged or fused by any makeshift statistical analogies.[1]

That it is not any special kind of experimental disturbance which leads to partially defined states, and the Uncertainty Principle, but rather the logical structure of quantum theory itself, has been pointed out by many authors.[2] Yet many continue to entertain 'counterinstances' to the general indeterminacy characteristic of the non-commutative algebra of EQT. Those who continue thus are reminiscent of *Through the Looking Glass*:

Alice laughed. 'There's no use trying', she said: 'One *can't* believe impossible things.'

'I daresay you haven't had much practice', said the [White] Queen. 'When I was your age, I always did it for half-an-hour a day. Why, sometimes I've believed as many as six impossible things before breakfast.' (p. 225, Dial Press (1931)).

It is to be hoped that these few bitter pills will have made 'breakfast' indigestible for a few philosophers of physics!

So, as Feyerabend once observed '...superstates cannot be incorporated into wave mechanics without leading to inconsistencies'. And as Condon had earlier put it: '...there is no place in the mathematical formalism of quantum mechanics for simultaneous exact position and momentum....'[3] I would wish to add only that superstates cannot be incorporated into any currently serviceable form of quantum mechanics without leading to inconsistencies. This, indeed, may be taken to constitute the heart of the Copenhagen Interpretation of Quantum Theory, just as it had constituted the heart of von Neumann's proof of the impossibility of hidden parameters functioning within *any* theory structured in accordance with the propositions *a* and *b* which open this chapter.

Feyerabend has written: '...the objective interpretation of the uncertainties as restrictive of the meaningful applicability of such functors as 'position', and 'momentum' will have to be adopted by any physical theory that incorporates the quantum postulate and the dual nature of (light and) matter....' [*Ibid.* p. 377.] In several essays[1] the present author has treated precisely this as the essential core of the Copenhagen Interpretation, with but two important additions: (1) past and present microphysical experience makes it probable (but in no sense necessary), that any future physical theory will incorporate the quantum postulate and the duality principle, and (2) there exists at present no wholly coherent, currently workable, and fully articulated conception of a microphysical theory which could do without this postulate and without this principle.

This much alone is all one needs to construe as the 'Copenhagen Interpretation'.

To that end I invite the reader again to distinguish the 'Copenhagen Interpretation' from the 'Bohr Interpretation'.[2] Writers like Feyerabend, Bohm and Vigier, on the other hand, construe the Copenhagen Interpretation as including all of the 'dogmatic elements' of the Bohr Interpretation. One need hold no brief for Bohr's naïve epistemology. Feyerabend is correct to score and underscore the strident statements of Bohr and Rosenfeld when they violate every lesson of the history of physics as they suggest that any future microphysics must, of logical necessity, 'guarantee' things like complementarity, irreducibly probabilistic laws, and quantum jumps. The very most a Copenhagen theorist can ever be in a reasonable position to argue is that:

(1) Any theory that incorporates the quantum postulate and the duality principle must also, if it would interpret the state-uncertainties objectively, restrict the meaningful applicability of functors like 'position', and 'momentum'.

(2) There are good contingent arguments in support of the expectation that any future microphysics will incorporate the quantum postulate and the duality principle—although no conclusive argument to this effect exists, or could exist.

(3) There exists now no genuine working alternative to the quantum theory we now have, notwithstanding all its awkward features. At most we have a variety of ingenious speculations on hand: not unworthwhile for that reason—but speculations none the less.

It is justifiable to joust against Bohr's metaphysical excesses. But I do not anticipate that many of Bohr's critics would be deeply troubled with points (1), (2) and (3) above, although they might express themselves differently concerning them. Yet, (1), (2) and (3) constitute the very framework of the Copenhagen Interpretation, while Bohr's metaphysics constitutes no necessary part of (1), (2) and (3). Extended conversations with physicists of all persuasions convince me that most of those who feel an affinity with the 'Copenhagen Interpretation' in virtue of (1), (2) and (3) above, feel no need whatever for Bohr's epistemology (ies).[1]

If the Copenhagen Interpretation is construed as Feyerabend, Bohm and Vigier construe it, then it should be fought at every turn. Philosophy of physics can ill afford to embrace dogmatism, even when endorsed by giants. But if the 'Bohr Interpretation' is cut away, then the Copenhagen Interpretation would seem to be eminently defensible. Or, does *die Feierabendglocke* continue to toll even for this 'liberalized' version of the Copenhagen Interpretation?

Here then is another short step towards our goal. Historically and conceptually the Copenhagen Interpretation and the Uncertainty Relations have been built into the very language of quantum theory—and it is only in terms of that language that the story of the positron concept can be told. The next chapter should provide us with the last few words we will need to tell that story.

CHAPTER VIII

EQUIVALENCE

A

Are Wave Mechanics and Matrix Mechanics really equivalent theories?

The writings of physicists so regularly and so strongly suggest an affirmative answer to this question that further inquiry might seem unprofitable. It will certainly strike some as a 'merely philosophical' query. However, the unguarded statement that Wave and Matrix Mechanics are equivalent physical theories is so unsound, historically and even conceptually, that a re-examination of the issue might still be tolerable. On our way to the positron, this is the shortest route.

First: a few boundary conditions of our inquiry. The assertion that Wave and Matrix Mechanics are equivalent theories is usually documented by the independent proofs of Eckart and Schrödinger (1926).[1] Consequently, our argument will initially contain a 'time cut-off' set at 1926.[2] Do the proofs offered by Eckart and Schrödinger really establish the theoretical equivalence of the Wave and Matrix Mechanics *of 1926*? That later systems—e.g. Dirac's Operator Calculus, or the general theory of Hilbert spaces—can generate solutions in either Wave or Matrix form (provided only that the state vectors amplify the partial differential equations), will not be relevant. By seeing just what was proved to be equivalent in 1926, a wider exploration of the philosophical aspects of equivalence may be facilitated, especially as these impinge on theories within physics.

Now for some initial conditions: it is a degenerate semantical point that for two physical theories to be equivalent they must be different—substantially so. If two theories differed not at all, they would not be equivalent, but identical. 'They' would be but one theory. The expressions '...are identical' and '...are equivalent' therefore do distinct semantical work. As physical theories, Schrödinger's Mechanics and Heisenberg's Mechanics could not have differed more. Their orientation and outlook diverge so radically that a 'proof' of their equivalence seems almost counter-intuitive. Indeed, how could any two genuinely different physical theories be 'proven' equivalent at all?

Consider this notion of equivalence a little further. When would we say two *notations* were equivalent? Again, they must differ substantially to begin with. Were they but typographically dissimilar, e.g. if one were set in Caslon face and the other in Janson, this would only be a single notation rendered in different printing types. Really different notations will differ with respect to the operations permissible in the formalism. Thus the system of logic based on 'Sheffer's stroke' differs from the first calculus in Whitehead and Russell's *Principia Mathematica*. But, though not identical, these notations are equivalent. They express the same theory—namely, the propositional logic. One criterion for judging non-identical notations equivalent, therefore, is that they express the same theory.

Consider now the equivalence of genuinely non-identical theories. On what criterion would this judgement of non-identity be based? *Ex hypothesi*, the theories are expressed in different notations; their formalisms are governed by different rules. What do we seek, then, when we seek to discover their theoretical equivalence?

Two (related) answers immediately come into prominence. The first is this: theories are equivalent when their postulate-sets, although (*a*) notationally, (*b*) psychologically and (*c*) sequentially dissimilar, turn out to be substantially the same sets.

(*a*) Notational dissimilarity was discussed a moment ago.

(*b*) Psychological dissimilarity consists in one theory merely seeming familiar and tractable while the other appears strangely unmanageable. Newton's theory of fluxions and Leibniz' differential calculus were certainly psychologically different. This is proved by the entirely different development these two versions of the calculus suffered at the hands of seventeenth- and eighteenth-century mathematicians. Doubtless, Schrödinger's and Heisenberg's Mechanics are dissimilar in at least this way too. Indeed most theoretical physicists think the two theories to differ *only* in this way.

(*c*) Sequential dissimilarity is nothing more than re-ordering of postulates. To place the parallel postulate at the very head of

EQUIVALENCE

one's geometry is to have a theory to that extent different from Euclid's.

What we are seeking to entertain here then is two theories whose postulates are ordered differently, governed by distinct formal rules, and responsible for different feelings, e.g. familiarity in one case against strangeness in the other. But now, despite all this, the postulates turn out to be substantially the same. Indeed, any two presentations of classical mechanics are likely to differ in just these ways. None the less, translation rules could easily transform any postulate of the one presentation into its analogue in the other. Would this situation then establish the equivalence of the two theories?

No. This is much too strong. That sets of apparently distinct postulates are really identical does not prove the equivalence of theories, but their absolute identity. Only the expositions would remain distinguishable—which is logically quite irrelevant.

The second characterization of equivalence is this: two theories are shown to be equivalent when it is established that every consequence of the one's postulate-set expresses precisely what some corresponding consequence of the other's postulate-set expresses. ['...every possible consequence' here means '...all well-formed theorems, whether or not they refer to observables'.] Thus, two theories are equivalent when one cannot construct an inference in one which is not a translation-analogue of some inference occurring in the other.

Again, this is too strong. It is not theoretic equivalence, but theoretic identity, which is thus proved. This characterization thus reduces to the first one set out above. Indeed, one way of proving apparently different postulate-sets to be identical (up to the stage of axiomatization) is to show that every theorem inferable from the one set is also inferable from the other. Conversely, to show that two theorem-sets are identical is to show that all members of each set follow from postulates which, though apparently different, are really identical. So we have here, in the above two characterizations, but one answer to our leading question. And this one answer is too strong; it delineates criteria for theoretic identity, not equivalence. By succeeding as a proof of identity it destroys those substantial differences which make two theories *two* theories. Two physical theories can never be identical. This generates again our earlier semantical point. One cannot both prove the equivalence of two theories *and* their identity; to establish the one rules the other one out.

Theoretic equivalence, therefore, is always '...with respect to subsets of the consequences of substantially different theories'. Wave and Matrix Mechanics (*c.* 1926), although often said to be equivalent in the strong sense, which reduces to an identity claim, are really equivalent only in this very limited sense. Indeed, this is the only sense of equivalence appropriate to physical theories.

Since physical theories are equivalent only with respect to subsets of their consequences, what kinds of 'defining' will determine 'a well defined subset'? Two different theories could be equivalent with respect to subsets of their *observational* consequences. These subsets might contain (*a*) all technically possible observational consequences, or (*b*) all physically possible observational consequences, or (*c*) all logically possible observational consequences.[1]

(*a*) *Equivalence with respect to all technically possible observational consequences.* Before the refined experiments of von Laue, Friedrich and Knipping (1912), the theory that X-rays were undulatory, and the alternative theory that they were corpuscular, were equivalent with respect to all technically possible observational consequences. Prior to this technical advance, wherein X-radiation was shot through a crystal made to serve as a diffraction grating, there was no experimentally available way of deciding whether X-rays were wave-like, or whether they were granular. Each theory made precisely the same claims concerning then-testable consequences. But as physical theories they were not identical, far from it. Hypotheses *more* opposed than that which argued for a particulate structure in X-rays, and another for a wave-like structure, would be hard to conceive. By extrapolation of just this point, Wave Mechanics *à la* Schrödinger, and Matrix Mechanics *à la* Heisenberg, could not themselves have been more opposed.

(*b*) *Equivalence with respect to all physically possible observational consequences.* Even before Eddington's eclipse observations (1919), it was clear that General Relativity was not equivalent—with respect to all *physically* possible 'observation-theorems'—to Newton's Celestial Mechanics. The former entailed that, during a solar eclipse, the fixed stars at the edge of the dark disk should be apparently displaced, due to 'spatial curvature' around so large a mass; or, rather, the different kind of space in that vicinity. Classical Mechanics entailed no such thing. As a matter of physical principle it could not have done so. Hence, concerning what it may be deemed physically possible to observe, nineteenth- and twentieth-century celestial theory are not at all equivalent. However, post-Einstein theories generally are equivalent to General Relativity in this particular respect, although they differ otherwise.

(c) *Equivalence with respect to all logically possible observational consequences.* Finally, two theories may be equivalent with respect to all *logically* possible observational consequences—although they may still differ concerning their non-observational consequences. The Gordon equation (1926) and the Dirac equation (1928) both had solutions, the infamous 'negative-energy' solutions, which were observationally uninterpretable until 1931, as we shall see. Hence they were, before 1931, non-observational consequences of the Gordon–Klein and Dirac theories. Thus, most physicists today would regard the existence of a *Perpetuum Mobile* (first type) as physically impossible, but not logically impossible. 'I've constructed a *perpetuum mobile*' is not demonstrably self-contradictory as is 'I've constructed a quadrilateral triangle'. It might be possible for two different theories to be equivalent with respect to all their logically possible observational consequences. This might be signalled by both theories ruling out physically impossible things like *perpetua mobiles*. But neither would convert this into an 'L-rule', or a formal property of the theory's syntax. Hence most versions of quantum theory (von Neumann's, Dirac's) treat the notion of commuting the position and momentum operators as constituting a syntactical impossibility: this is tantamount to being logically impossible within systems so defined. Bohm, on the other hand, and Vigier and Feyerabend treat this non-commutativity as constituting an experimental limitation only. Hence Bohm's 'theory' is not equivalent to Dirac's with respect to all logically possible 'observation-theorems.'

Here then is the crux. How can equivalence in any of these senses, (a), (b) or (c), ever be established? In so far as equivalence between the physical theories is usually 'with respect to observational consequences' (characterized as above), there can never be general proofs of theoretic equivalence.[1] Consider any two theories, T_1 and T_2. What is done to show their equivalence? We take one observation statement from T_1 and match it to a corresponding statement in T_2. If they make the same claim, then T_1 and T_2 are equivalent with respect to that consequence. Suppose consequences $x, y, z,$ of T_1 are matched against consequences x', y' and z' of T_2. If all these refer to what is technically possible to observe, then one can say that (with respect to these three 'technically possible' observational consequences) T_1 and T_2 are in so far equivalent. But, that T_1 and T_2 are equivalent with respect to *all* technically possible observational consequences, this claim would be an inductive generalization based on too few correlated instances (in this case, three).

Whenever we claim the equivalence of T_1 and T_2—irrespective of whether this pertains to what is technically, physically, or logically possible to observe—this is always an *inductive* claim. It consists in comparing representative consequences of T_1 and analogous consequences of T_2. Since the consequences of any well formed theory constitute a potential infinitude, 'proof of the equivalence of T_1 and T_2' can never be stronger than 'well confirmed hypothesis concerning the consequential isomorphism of T_1 and T_2'. In so far as we are discussing physical theories, and not simply their formal calculi, any 'proof' stronger than an inductive proof is ruled out as a matter of principle.

Consider yet further the idea of a physical theory. This will reinforce some distinctions just set out, and introduce our discussion of early quantum theory in such a way as to focus upon the positron discovery.

A physical theory is, at least, a contingently interpreted formalism—a delicate trinity of algorithm, physical interpretation and correspondence rules.[2]

A single algorithm, given two quite different physical interpretations, results in two quite different physical theories. Suppose *you* co-ordinate Euclid's 'straight line' with observed pencils of light. *I* co-ordinate it with observations of taut wires—or the surfaces of optically flat glass—or even the trajectories of freely falling terrestrial bodies. We thus concern ourselves with different physical theories, although our algebraic computations may be indistinguishable. We will hunt for different things in testing our theories.

Our criteria for precision in measurement will differ. The avenues along which research will seem promising are not likely to be the same for us both.

In contrast, two quite different algorithms, given the same physical interpretation, may result in equivalent physical theories; 'equivalent, with respect to suitably defined subsets of their consequences'. The algorithms of geocentric and heliocentric astronomy differed structurally. They were managed by different rules, were anchored to different reference points and contained by different boundary conditions, not to mention entirely different general ideas concerning which operations would be appropriate within the algorithm. But, ignoring stellar parallax, for now, the Ptolemaic and Copernican astronomies were equivalent for most pre-Galilean 'naked eye' observations. Concerning the technically possible observations of the fifteenth century, both theories made the same claims. In so far they were equivalent. Both theories construed the term 'stars' (used also to designate planets, as in Kepler's 'De Motibus Stellae Martis') so as to denote little more than points of light moving inside a black sphere of unknown radius. Before the advent of the telescope, therefore, this common physical interpretation generated from each theory the same predictions. This made the two theories (in so far) equivalent.[1]

So, on this account, a single algorithm variously interpreted can result in quite different physical theories, while different algorithms, similarly interpreted, can result in equivalent physical theories (with respect to subsets of their observation statements).

As a physical theory, the Wave Mechanics of 1926 was the elegant Schrödinger algorithm *plus* the Schrödinger interpretation. Matrix Mechanics (in 1926) was the Heisenberg–Jordan–Born algorithm *plus* a corresponding interpretation.[2] These two physical theories never could be proven equivalent; they are *fundamentally* different.

What then do the Schrödinger and Eckart papers purport to prove? Nothing more than the mathematical identity of Wave and Matrix Mechanics. Their expositions do not stress clearly enough that this is not the same as proving the equivalence *as physical theories* of Wave and Matrix Mechanics. On this most commentaries err. Perhaps this is because Eckart and Schrödinger were not themselves pellucidly clear about their own procedures. To many investigators it has seemed that if two theories are *mathe-matically* identical—itself a very strong claim—it follows that as physical theories they will be equivalent.[1] Our earlier arguments were designed to throw doubt on this point.

What the Eckart and Schrödinger proofs may be taken to do, at least at this stage of our inquiry, is to show that the calculi of Wave and of Matrix Mechanics are intertranslatable. There is no theorem generable in T_1 which lacks an analogue generable in T_2; more formally, there is no claim which can be made in the one algorithm which cannot correspondingly be made in the other—such that, which version of quantum mechanics we use becomes a matter of preference, taste, or other pragmatic and psychological factors.

B

Most contemporary physicists would argue that no problem can be formulated matrix mechanically which cannot also be formulated wave mechanically—with identical observable results. After all, it is argued, the relation between the two formulations is hardly more than the relation between two reference frames. Any state [of motion, or of variation in any variables] can be represented as a vector in an abstract mathematical space (fig. 1):

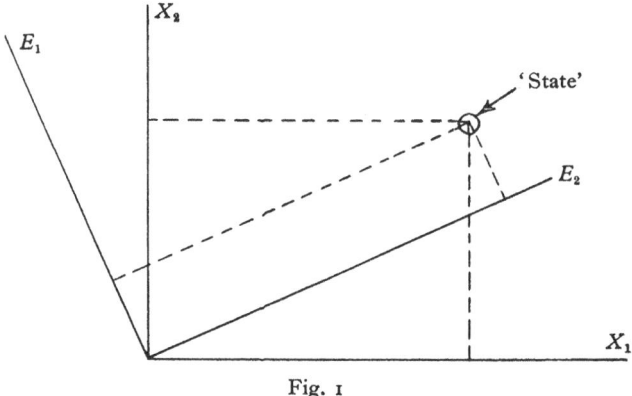

Fig. 1

Matrix Mechanics resolves that vector on 'axes' of definite *energy*; certainly the original Heisenberg Mechanics does this. Wave Mechanics chooses resolutions on axes of definite *positions*. Dirac's calculus manages without any resolutions whatever—

dealing directly with the state vectors themselves, rather than with their components upon any axes. At present, physicists switch back and forth from one approach to the other; any distinctions between Matrix and Wave Mechanics are now virtually forgotten. By the employment of some familiar algebra this almost trivial equivalence between Wave and Matrix Mechanics can be made apparent (see appendix III). Because of the possibility of such derivations, contemporary physicists are inclined to treat Wave and Matrix Mechanics as equivalent ways of expressing precisely the same thing. Moreover, some would treat all this as having been implicit in the original formulations of Wave and Matrix Mechanics by Schrödinger and Heisenberg. About this views may vary considerably. What is involved are different definitions of 'physical theory'. It might be argued that, at any time, a physical theory is no more than the totality of all its explicit developments for specific problems up to that date. Or, perhaps all of the developments implicit in it are also 'part' of the physical theory. This last approach will treat theories as 'operationally defined' or 'open-ended'; their complete 'definitions' are never finished so long as there are untried applications in prospect. Thus, for example, there may have been problems in 1926 to which Matrix Mechanics had already been applied, and hence developed, while Wave Mechanics had not yet at that time been viewed in such connexions. Some physicists doubt even this. Wave Mechanics was more energetically developed, since it allowed the use of familiar mathematical techniques. But it may be argued that every quantum mechanical problem could in principle have been done by either technique even in 1926, because a mathematical identity of the two theories was then implicit in the algorithms.

Still, one can take the stricter view. Granted, it is possible to treat classical mechanics as a single organic theory all of whose present applications were implicit in 1687. But it is also possible to refer to 'Newton's physical theory', meaning just what was published in 1687, making no covert reference to the later developments of that science. Indeed it would seem to me simply to muddle the history of physics to do otherwise. Similarly, it is possible and legitimate to contrast Wave Mechanics of 1926 with Matrix Mechanics of 1926, and to ask whether—then—there were grounds for claiming either the mathematical identity, or the equivalence as physical theories, of these works of Schrödinger and Heisenberg. This question can and should be viewed strictly from the point of view of 1926. It need not be compounded with the hindsight arising from preoccupation with later developments. The philosopher and historian of science occasionally must so view events in the evolution of ideas. To understand the threshold-structure of quantum theory as Anderson, Dirac and Blackett moved through it towards the positron concept it is necessary for us so to view the events before us.

At the time of the Schrödinger–Eckart 'proofs', the relevant theory of Hilbert spaces was not known. Hence no one could have done more than to show a formal analogy between Wave and Matrix Mechanics.

The matrix theory was itself a direct outgrowth of the work of Bohr and Kramers on atomic energy levels and on the intensities of spectral lines. At that period the major interest of all atomic physicists lay in the calculation of atomic and molecular energy levels and in the classification of spectral series. The great emphasis on scattering problems, which is characteristic of the present epoch in quantum physics, had not yet arrived.

The theory used by Bohr and Kramers was based on the classical theory of multiply periodic mechanical systems. For such systems the co-ordinates of the particles, in the classical sense of Newtonian mechanics, are expressed by Fourier series with a set of different frequencies and all of their overtones. Lissajou's figures in the plane give a common example of this treatment. Following the arguments of Bohr's Correspondence Principle (cf. chapter IV), the magnitudes of the coefficients in these series were related to the specific intensities of the individual spectral lines. In this way at least the relative intensities of the lines in some spectral series could be calculated in a reasonable and satisfying manner. The theory as thus described grew out of the much older idea that the spectral lines were to be uniquely associated with particular frequencies of motions of the particles.

While this method enjoyed great success in practice, its physical meaning was rather obscure. It was found also that the method (and its constituent techniques) was not entirely straightforward; some rough and *ad hoc* manipulation of the results had to be resorted to at times to satisfy such conditions as the sum rules for the intensities. Heisenberg undertook to rationalize the basis of the theory by abandoning the idea of there being any intimate connexion between Fourier series and spectral intensities. He attempted to replace the calculations with series by more formal algebraic rules which would work with observable frequencies and intensities. This 'new kinematics', as it was called, was ultimately recognized as virtually constituting a matrix calculus. From this point on the details of the matrix mechanics were established with admirable insight and remarkable alacrity by Born, Heisenberg and Jordan. An equivalent form of operator calculus was established a little later in this same period by Dirac.

The invention of Wave Mechanics was a wholly independent event. In an arresting piece of reasoning by analogy, already described, Louis de Broglie argued that if electromagnetic radiation could manifest the properties of both waves and particles (wave-particle duality), the same might be so for electrons (particle-wave duality). He tried unsuccessfully to develop a theory in which the waves could guide the particles, in some sense, while the latter would be identified with a kind of singular region in the wave system.

Schrödinger was inspired by De Broglie's work to try his own hand at the problem. He treated the connexion between waves and particles in a much looser manner than had De Broglie; quite soon he discovered the differential equation which has since then gone by his name. From here on his progress was very rapid, and in a few months he had delineated virtually the complete mathematical apparatus of non-relativistic wave mechanics as we know it today.

It is remarkable that both forms of quantum theory were designed, by formal mathematical methods, before there existed any substantial physical interpretation of the new procedures. The statistical interpretation, as suggested by Born and amplified by von Neumann, and the theory of measurement as developed by Heisenberg and by Bohr, are later additions to an already accepted mathematical theory. Surely philosophers and historians of science should find this a matter of great interest! It makes our earlier references to Mr Notwen, in chapter II, somewhat less fanciful than they may have appeared there.

Up to this stage (1926) there had been no inkling whatever that

Matrix and Wave Mechanics were related in any significant manner, even though it was obvious that the two theories were yielding 'equivalent' numerical results for most practical problems. The physical backgrounds of the two theories, as well as their mathematical techniques, were so vastly different that it was legitimate to doubt that they could agree for all problems. Most physicists did so doubt. The papers by Schrödinger and by Eckart, in which the existence of a formal connexion between the theories was demonstrated, hit physicists with the force of revelation. It is but natural that in these circumstances the actual and justifiable significance of the revelation may have been slightly misinterpreted.

This conceptual drama was brought to its climax with the later development of a unified theory of quantum mechanics by von Neumann.[1] His great work led many mathematicians and physicists to the easy conclusion that the proof of equivalence of all forms of the theory was now complete. Unfortunately, the facts have never justified this degree of optimism.

Von Neumann's theory was a splendid achievement; but it was also a precisely defined mathematical model, based on certain arbitrary, but very clearly stated assumptions concerning quantum theory and its physical interpretation. The plain and unvarnished fact is that physicists have made almost no direct use of von Neumann's methods in practice, nor have they bothered to try to stay within the bounds which he set.[2] Only recently have serious and sustained efforts been made to see whether, in fact, it is possible to formulate scattering problems within the limits of his elegant theory.[3] Some definite progress has been made, but it has not yet been possible to show, for example, that the important problem of scattering in the Coulomb field can be solved in anything like this way.[4] The methods and techniques used in quantum field theories fall almost wholly outside the framework of von Neumann's theory.[5]

Even within the bounds of von Neumann's formulation the Matrix and Wave Mechanical versions are not strictly equivalent. When a problem requires the use of continuous spectra the matrix method just cannot be fully applied. One of Heisenberg's original assumptions was that the energy of a system should always be representable by a diagonal matrix. When there is a continuous spectrum this diagonalization process simply cannot be carried out, and a much more detailed theory of the spectrum is clearly needed. It is one of the great mathematical beauties of von Neumann's theory that it gives a unified approach to both discrete and continuous spectra. The mathematical formulation of quantum theory by methods which are closer to those used by physicists than is the technique of Hilbert space analysis given by von Neumann has been given recently by Titchmarsh.[1]

There remain many unsolved problems in the region of quantum theory for which both Matrix and Wave mechanical methods are applicable. The energy level systems of atoms, molecules and nuclei seem to be distinctly simpler in structure than is contemplated in von Neumann's abstract theory. It is possible that the use of group-theoretical considerations will lead to a clearer picture of bound energy states than we have at the present time. On the other hand, the mathematical difficulties associated with the quantized field theories are clearly of much deeper logical origin than those of non-relativistic theory. It is probable that quantum theory will need to undergo fundamental revisions before those problems can be fully solved. But, as we have been urging through these many pages, such revisions will be fully effected only by some stiff algebra and penetrating physical insight, and not simply by speculations about determinism, about possible hidden entities and the 'lessons' which ought to have been taught us by the history of physics.

Let us return to the specifically historical question concerning the 1926 equivalence of Wave and Matrix Mechanics; the following problem may now be relevant.

Let E be an electron free to move in one direction (one-dimensional space). Let q be the position-co-ordinate, and let the 'eigenfunctions' of position be ψ_1, ψ_2, ... For simplicity's sake assume a discrete position spectrum.[2]

Now, what corresponds to the 'position' of E in the matrix theory is a q-matrix $|q_{ij}|$. The theory determines completely the values of the entries q_{ij} in the case that the matrix has been diagonalized (by making a position measurement). But in the general case of an 'unsharp' position, the theory determines only certain relations between the q_{ij}—and these are never sufficient to compute the q_{ij} themselves.

The usual 'correspondence' between matrix theory and wave theory involves the following rule for determining the q_{ij}:

$$q_{ij} = \int \psi_{i(r)} q \psi_j \, dx. \qquad (1)$$

The point to be raised here is that (1) is independent of the matrix theory—certainly as it stood in 1926.[1] If (1) is accepted, then the two theories (or at least their algorithms) do indeed coalesce—and are most intimately related from a formal point of view. But accepting (1) involves rejecting certain (so-called) 'realistic' interpretations of Matrix Mechanics, and these are, moreover, among the most attractive.

The crux is that this rule is independent of Matrix Mechanics. If so, Matrix Mechanics involves variables of potential physical significance which are not present in Wave Mechanics (the q_{ij}). And (1) in effect robs the q_{ij} of any real significance by saying, for example, that the q_{ij} depend only on the ψ_i. Otherwise, two different electrons with the same set of eigenfunctions might have different matrices, although the matrices would, of course, have to be equivalent under diagonalization.

This example reflects in a specific way the general issues we have discussed concerning the identity, or the equivalence, of Matrix and Wave Mechanics (1926). However, it is not our primary objective to appraise the formal validity of the early proofs of Eckart and Schrödinger. Even if we assume that these early proofs are formally valid, we are still left with a philosophical problem— that concerning in what sense Wave Mechanics and Matrix Mechanics may be considered equivalent *physical theories*.

C

Assume then that the Eckart–Schrödinger proofs of mathematical identity are valid; that the Operator Calculus does mechanically transform every Wave mechanical statement into a corresponding Matrix mechanical statement.

This much alone would never prove the identity or even the equivalence of Wave and Matrix Mechanics as *physical theories*. The only way that this could be achieved would be to establish both (1) that but one algorithm served both theories, and (2) that the same physical interpretation was in each case pressed on to the algorithm. Here, in the present case, neither (1) nor (2) seems to have been achieved. Because, although the Eckart–Schrödinger proofs may establish that Wave and Matrix Mechanics are intertranslatable (via the Operator Calculus) this alone does not prove that only one algorithm is used for both. Just because there is a transformation language for converting every statement of geocentric astronomy into a corresponding heliocentric statement, this by itself does not render the original algorithms into one indivisible calculus.

But, for the sake of the present argument, let us treat this last objection as unnecessary. Let us suppose *pro tem* that T_1 and T_2 are mathematically identical; ultimately, one algorithm could serve both. Item (2) still remains unfulfilled.[1]

Historically, the physical interpretations of Wave and Matrix Mechanics could not possibly have differed more. Here a possible heresy on my part may be detected. The infamous Schrödinger interpretation is to be treated in this argument as a genuine (although observationally false) physical interpretation of an algorithm. Too many physicists have treated the early continuum interpretation of Schrödinger as 'mere metaphysics', while the probabilistic interpretation of the ψ function has been supposed to be implicit all along in the Matrix formulation, becoming explicit finally in late 1926 in Born's great paper. This has encouraged commentators to suppose that the Eckart–Schrödinger 'proofs' achieved more than they could have done. It is supposed that the equivalence of two physical theories was proved, rather than just a formal identity between algorithms (assuming still that this *was* proved).

No. The Schrödinger interpretation of the ψ function was a *physical* interpretation. It was not just metaphysics; Eckart and Schrödinger even anticipated experiments which might ultimately decide in favour of a continuum theory rather than a statistical, discontinuous account of microparticles.[2] Schrödinger always held that the phase waves (in configuration space) represented observables. Proponents of the continuous substratum theory (Schrödinger and De Broglie) never supposed a description of an 'observed' phase wave could be given here and now. But they took this to be but a technical limitation.

Born demonstrated that *physically* existing phase waves would require experimental spaces of $3N$ dimensions (N = the number of particles). Thus nine actual dimensions are needed in any three-body problem. This destroyed the plausibility of Schrödinger's interpretation by disclosing that it entailed statements which were either demonstrably false or physically unintelligible. Still, a continuum theorist might stand his ground, suggesting that just as we have no idea of what an observation of phase waves would be like, so also we have no notion as to how the expression 'spatial dimensions' might require re-interpretation for the understanding of new, and unprecedented, phenomena. But it is enough here to treat the Schrödinger interpretation so that it generates some observation statements which are empirically false. From which it follows that the interpretation cannot be 'merely metaphysical'. Merely metaphysical theories are not even empirically false; the Schrödinger view at least has this merit. The fundamental, material substratum, whose effects were manifested in the observation statements of Schrödinger's theory, was understood to be non-granular, non-particulate and non-discontinuous. Roots of this view run through De Broglie, Clerk Maxwell, Hamilton, back to Fermat. Quantum phenomena, point scintillations and other apparent discontinuities, were but the surface manifestations of more fundamental field-like phenomena.

Heisenberg's original departure was completely opposed to this. His strict adherence to observables determined that discontinuity would be, from the beginning, the conceptual background of Matrix Mechanics. Matrices were always construed as the mathematical counterpart of actual scattering patterns of particle populations, or as revealing other physical properties of particles *en masse*.[1]

Assuming now that the Schrödinger–Eckart proofs established some formal identity between the Wave and Matrix algorithms, what further was required to show the equivalence of these *as physical theories*? Born gives an answer. Since Schrödinger Wave Mechanics, and Heisenberg Matrix Mechanics, were born with different interpretations built into them, any attempt to stress their similarities would seem to require forcing the interpretation-question far underground. The actual proof of their equivalence as physical theories was achieved not by Eckart or Schrödinger, but by Born himself. By pressing to the limit the statistical interpretation of $|\psi_{(Q)}|^2$ he infused both algorithms with the *same* physical interpretation. Assuming them to be mathematically identical, and by simply grafting on them the same interpretation, he rendered them equivalent, as physical theories, with respect to all technically, physically, and logically possible observational statements. But in doing so he drastically changed the physical theories which had originally gone by the names 'Schrödinger Mechanics' (T_1) and 'Heisenberg Mechanics' (T_2). He replaced Schrödinger's continuum interpretation of the ψ function with a statistical account; and he made explicit what had before been merely implicit in Matrix Mechanics, namely, that the microphysicist could concern himself not with singularities, but only with distributions and statistical ensembles of particles. That is, Born proved not the equivalence as physical theories of T_1 and T_2, but of two quite different physical theories, T_3 and T_4. Now T_3 and T_4 were beautifully made for subsumption within some more general theory, e.g. that of Dirac, or of von Neumann. But it was these Born-modified physical theories which became components within the larger theoretical frameworks of the 1930's, and not those originally advanced by Schrödinger as Wave Mechanics (T_1), and by Heisenberg as Matrix Mechanics (T_2). By finding the Matrix mechanical analogue of $|\psi_{(q)}|^2$ and by interpreting both that matrix and the ψ function as giving the probability of finding a microparticle within a given volume element,[1] Born alone deserves the credit for establishing the equivalence as *physical theories* of Wave Mechanics (i.e. T_3) and Matrix Mechanics (i.e. T_4). Historically speaking, it would be erroneous to suppose that anything approaching Born's achievement is to be found in the Eckart and Schrödinger papers.

Max Born, then, supplied the only really convincing argument of the physical equivalence of Wave Mechanics (T_3) and Matrix Mechanics (T_4). But did he prove this equivalence? Born's argument is entirely inductive in nature. What he shows is this: that for every Born-interpretable observation statement within T_3 there is in fact a corresponding Born-interpretable statement within T_4. Even when T_3 and T_4 are imbedded within a more comprehensive theory, e.g. the elegant Operator theory of Dirac, Born's thesis holds for every problem *so far considered*—the hydrogen atom, Compton scattering, Zeeman effects, electron spin, radiation effects, harmonic oscillations, ..., etc. This creates a very strong inductive presumption that for *any* problem formulable within T_3 (i.e. the Wave equation development of the Dirac theory), there will be a corresponding formulation in T_4 (i.e. the Matrix aspect of the Dirac theory). But there is no general proof that this will always obtain. Nor could there be. Not so long as we are studying physical theories (which are 'open-ended' concerning future possibilities), as opposed to uninterpreted algebraic formalisms. Consider again the Eckart and Schrödinger proofs themselves. Imagine Wave Mechanics and Matrix Mechanics as but purely uninterpreted calculi. Are they really formally identical? Do Eckart and Schrödinger actually prove this?

Let us reconsider just what Eckart and Schrödinger achieved. Consider now two calculi, C_1 and C_2. These are proved to be the *same* calculus when it is demonstrable that C_1 and C_2 have the same postulate-sets (and the same transformation rules). Or (what is the same) that the total consequence-set of C_1 is both isomorphic and structurally analogous with the total consequence-set of C_2. Either procedure would constitute a rigorous proof of identity between two apparently different formalisms; these proofs would be achieved by the use of logical transformations alone.

Now, is it the case that the Operator calculus is nothing more than a set of logico-algebraic transformations of a high order of generality? Begin by assuming this.

A proof showing C_1 and C_2 to be mathematically identical would consist not only in the provision of logical transformations converting any selected statement of C_1 into a corresponding statement of C_2. It would also establish that for all *possible* statements within C_1 there is a transformation into a corresponding statement of C_2. Do Eckart and Schrödinger provide such a proof? Certainly not. What they do do is select 'typical' or 'paradigmatic' types of microphysical problem (e.g. the harmonic oscillator, Compton scattering, Doublet atoms, etc.), and show that, via the Operator calculus, every Matrix mechanical formulation and solution of these problems has some Wave mechanical analogue. But this is not one whit less inductive than the Born 'proof'. Neither Eckart nor Schrödinger bar against the possibility of some as-yet un-derived theorem of Matrix Mechanics *failing* to translate (via the Operator calculus) into a corresponding statement of Wave Mechanics—or vice versa. Although they purport to prove the identity of C_1 and C_2, Eckart and Schrödinger adopt a procedure which within any arbitrary set of mathematicians would be suspect, and which would never constitute an identity-proof for two mathematical systems. Similarly, one could never hope to prove that an even number is always representable as the sum of two primes by running serially through the even numbers, and decomposing each into primes; neither can one prove mathematical identity in any such an inductive manner. Of course, it may appear inconceivable that any particular wave equation should not translate into matrix form. But this psychological confidence does not constitute even the beginning of a proof; for all I know there may actually *be* a rigorous proof of the identity of Schrödinger's Mechanics and Heisenberg's. But neither Eckart nor Schrödinger provided that proof, since they did not show and could not show that their inductive procedure was logically exhaustive.

Yet one further difficulty plagues these identity-proofs. Suppose we construe the Operator calculus not simply as a logical transformer, or an elaborate translation rule, but rather as a well developed algorithm in its own right. Physicists have done precisely this. Then, presumably, the Eckart–Schrödinger situation could be represented somewhat as follows: from the comprehensive Operator calculus one can generate (as formal consequences) the subcalculi called 'Wave Mechanics' and 'Matrix Mechanics', just as classical optics and early electro-dynamics can be regarded as independent consequence-sets of the more comprehensive, unified theory of Clerk Maxwell. That is, the whole calculus of Wave Mechanics (C_1), and that of Matrix Mechanics (C_2), can be viewed as consequences of the inclusive Operator calculus (O). Thus it immediately appears that any statement within C_1 could be generalized and transformed through O into a corresponding statement within C_2. Several commentators have characterized the Dirac Operator calculus in this way. A general theory, enriched with the apparatus of Hilbert spaces and state vectors, thus generates with uniform facility any statement of C_1, or any statement of C_2. Within a calculus so rich it becomes a matter of logical indifference which manner of expression one chooses (C_1 or C_2). From this fact alone it has sometimes been assumed that therefore C_1 and C_2 are formally identical. Indeed, this inference is made with increasing frequency at the present time. But this is logically objectionable. From the fact that O implies C_1, and the quite independent fact that O implies C_2, it does *not* follow that C_1 implies C_2. This would be like inferring from the fact that 'being a bachelor' implies both 'being a male' and also 'being unmarried', to the conclusion that 'being a male' implies 'being unmarried'. This is demonstrably unsound.

Thus there are two distinct possibilities *vis-à-vis* the status of the Operator calculus. It is either a collection of logical transformers, lacking all semantical content, designed to convert C_1 into C_2 or vice versa—or else it is itself a more generalized algorithm, with respect to which C_1 and C_2 are the deducible consequences. This second alternative is to be rejected as a proof of the identity of C_1 and C_2, for the argument just given. The first account, however, seems in some respects inconsistent with the procedures actually adopted in the Eckart–Schrödinger papers, which procedures are most properly characterized as 'inductive' in nature.

Born's 'proof' was about as rigorous as any demonstration of equivalence between physical theories could be. Such a demonstration as his could never be any more than an inductive comparison of observation statements in T_3 with those in T_4. Consider these facts: (1) T_3 and T_4 are both given substantially the same physical interpretation (i.e. Born's statistical rendition of the ψ function). (2) Whatever their other limitations, the Eckart–Schrödinger proofs profoundly suggest some formal intimacy to obtain between C_1 and C_2. (3) No *substantial* objections to the efficacy of Born's interpretation have ever been raised, at least not by experimentalists. From all this, (1), (2) and (3), we would seem to have the strongest demonstration of equivalence that one could ever have between physical theories. It is no *proof* in the strict mathematical sense, however.

But it is a conjecture stoutly to be confirmed.

The Eckart–Schrödinger *proofs*, however, suggest mathematical rigour of a different order. They suggest that the idea of a supposed theorem of C_1 which could not be transformed (via O) into a corresponding statement of C_2 is somehow demonstrably untenable. And so it would be, too, if their proofs had the formal rigour their authors suggest. But for this goal, nothing less than a *general* mathematical demonstration will suffice. The wholly arbitrary selection of typical solutions within C_1 which are then singly 'mapped' into corresponding solutions of C_2, although this *suggests* the existence of the proof in question, does not itself constitute that proof.

In sum: writers on physics, and this includes historians and philosophers, as well as physicists, have characterized the Eckart and Schrödinger proofs of 1926 as establishing the equivalence of Wave Mechanics and Matrix Mechanics as physical theories. Our objective here has been to show that these 'proofs' establish nothing of the kind. Indeed, although Eckart and Schrödinger could have been a lot clearer on this point, they do seem to be aware that their 'proofs' have to do only with the formal identity of the calculi C_1 and C_2. Do they actually establish this mathematical identity? As we have seen, two views are possible: (1) They subsumed Wave Mechanics (C_1) and Matrix Mechanics (C_2) within a more comprehensive theory—the Operator Calculus (O). From this it does not follow that C_1 and C_2 are identical. The identity of two theorems or sets of theorems could never be established just by their following from the same premiss or set of premisses. (2) O is nothing more than a logical transformer, and completely lacking in any semantical content, which converts any statement of C_1 into a corresponding statement of C_2. This it does without adding anything of its own. But that this is the case could never be proved by an inductive selection of typical problems within microphysics; the actual procedure of Eckart and Schrödinger. Strictly speaking, it could be consistent to doubt that C_1 and C_2 are ultimately identical even after entertaining such a 'proof'. Finally, the really convincing argument for the equivalence *as physical theories* of Wave Mechanics and Matrix Mechanics was established by Born's statistical interpretation of the ψ function. Because here, in an openly inductive procedure, Born forces a common physical interpretation on to both C_1 and C_2. This makes it a matter of indifference which algorithm one chooses in which to express his predictions.

This has been our steepest step yet towards the positron concept. But now we should be ready. The conceptual context within which Dirac's work in 1928 and Anderson's and Blackett's in 1932–33 was set has been carefully delineated. Born's interpretation, being itself built into the Dirac calculus, set the stage for the great drama of ideas within which the discovery of the positron was the most exciting moment. Nothing less than the entire theory of anti-matter—with its attendant subtheories of 'materialization' (of matter out of energy) and of pair-creation and annihilation—was the ultimate consequence of this 'single' discovery. The degree to which the theoretical efforts of Schrödinger and Heisenberg formed part of this discovery is now before us, as is the central role of the Copenhagen Interpretation during all these critical years. The structural and formative function of the Uncertainty Relations within the Anderson–Dirac–Blackett enterprise has also been set out. An historically more acceptable analysis of the Corresponding Principle has been delineated, and this will certainly help us over two or three rough places in the next chapter—the final one. General philosophical morals were earlier drawn about the functions of pictures and models in microphysics, about the logic of the properties and predicates we ascribe to micro-entities, about the nature of explanation and prediction within quantum physics, and about the theoretical role of crucial experiments throughout the history of modern physics. If all these steps have been designed properly, our final chapter should fall into a pattern all parts of which will be intelligible because they have been adequately prepared for. The reader alone will be able to judge whether or not the author's architectural plan has been sound.

CHAPTER IX

THE POSITRON

*On 2 August 1932, Dr Carl D. Anderson
discovered the positive electron*

This remark is often repeated in textbooks and manuals on modern physics. It is singled out here for several reasons: (1) its innocent, matter-of-fact tone conceals one of the most intricate and interesting chapters in the history of scientific discovery; (2) it encourages an analogy between discoveries within microphysics, and quite noncomparable discoveries within, for example, natural history; and (3) the actual discovery of the positron, in all its dimensions, constitutes an instructive example of the interplay of theory and experiment within physical science, one of the best a historian, or logician, of science could ever hope for. It also constitutes the denouement of this book.

The positron packet can be dipped into in three different ways. The physicist reaches into this complex of concepts to facilitate his thinking within, for example, quantum field theory, or the experimental study of cosmic rays. The historian of physics attends, as we have been doing, to the exciting interplay of ideas within the microphysics of the 1920's—the better to perceive the initial resistance to the positron-idea, as well as to trace the theoretical evolution which resulted in the claim that positive electrons exist. The logician will focus upon the internal structure of the arguments and concepts actually employed, and still being employed, by those physicists who have played major roles in this story. We have been already doing this with related concepts. In the history of science the positron discovery constitutes at once one of the richest, and yet one of the most neglected, areas of inquiry.

The discovery of the positive electron was a discovery of three different particles. I choose to put the matter this way, rather than to remark three different discoveries of the same particle, because the conceptual backgrounds within which the work of Dirac, Anderson and Blackett took place were so disparate as to leave it unclear until almost 1934 whether their findings had anything in common. Perhaps this is not so dramatically apparent in the case of Anderson and Blackett, since the latter knew immediately that 'his' particle was the same as Anderson's. Dirac's work is quite distinct, however.

Indeed, a contemporary reflexion of this is found in the ways in which scientists describe the discovery. Thus Millikan, writing in 1935, says '...the discovery of the positive electron...was made without the guidance of any theory whatever, as was also the discovery of the frequent occurrence of positive-negative pairs of tracks...'.[1] Professor Blackett describes the discovery in much the same way, as does also Dr Oppenheimer and Anderson himself.[2] On the other hand, Professor Hans Bethe feels the discovery to have been primarily a theoretical one, on the grounds that positron tracks had been photographed before 1932. However, lacking a theory in virtue of which such tracks might be made intelligible, physicists failed to identify them. Indeed, Professor Konopinski avers that 'every theoretical physicist knows that Dirac discovered the positron'. Thus, even today, these historically different approaches to the particle remain unresolved.

Accordingly, this chapter will be designed as a triptych; we shall discuss (A) the Anderson Particle, (B) The Dirac Particle, and (D) The Blackett–Occhialini Particle. [Section (C) is devoted to a closely related issue.]

Even before considering Anderson's research, some account of the genesis of the present chapter may be in order. Professor P. A. M. Dirac once spoke to me of a lecture given at the Cavendish by D. Skobeltzyn, 'sometime in 1926 or 1927'. Dirac recalls the description by someone, perhaps Skobeltzyn, perhaps someone in the audience, of an experimental setup within which Skobeltzyn was bombarding a metal target. One of the curiosities reportedly mentioned by Skobeltzyn was that several particles which were certainly electrons were seen to 'fall back into the source'; this, despite the fact that most of the electrons moved in the way usual for this experiment, *away* from the source. Professor Dirac feels that what he remembers Skobeltzyn as having then described could only have been positive electrons, and he suggests that the Russian might very well then have made the discovery.

My reaction was the normal one for a philosopher or historian of science, that it would have been an immensely difficult feat of

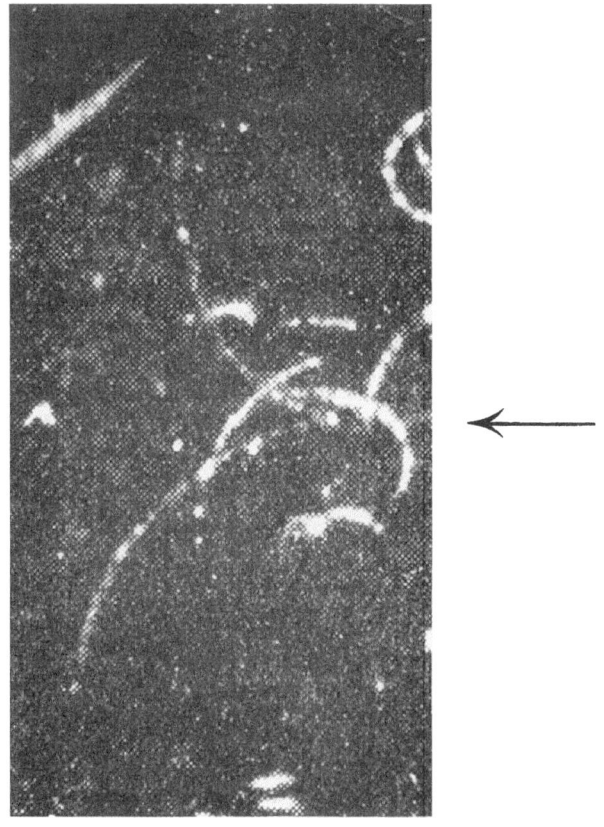

From *Zeitschrift für Physik,* 1927, XLIII, p. 362.

imagination for Skobeltzyn, or anyone else, to have made this discovery in 1926, given the theoretical conceptions and experimental facilities then available. An exploration of Skobeltzyn's published work entirely failed to reveal any photograph, or even a passing remark, relevant to Dirac's narrative. However, a related photograph did turn up in Z. *Phys.* XLIII (1927), 362. Schematically reproduced in fig. 2 is part of the lower left-hand photograph on that page.

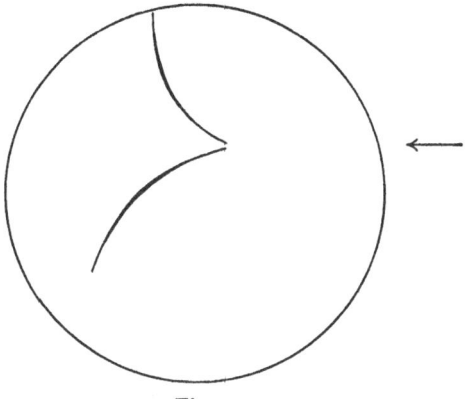

Fig. 2

Let me stress that this configuration is probably not what it appears to be. None the less, the presence of geometrically associated tracks, of the sort one expects with positron-negatron pairs, will be evident. As a check on my own reactions this photograph was shown to a great many experimentalists. Without exception, all granted that this might well be a pair; some thought it a very good example of one. No stereoscope was used in this informal survey, however.

Professor Skobeltzyn, in two communications of 1956 and 1960, offers a wholly different interpretation of this photograph (see Appendix IV). He takes it to be a chance overlapping of two different tracks, one of a Compton recoil-electron, and the other of a spurious track emerging from the chamber wall. Without doubt, Skobeltzyn is correct, but since the original plates are destroyed, the use of a Pulfrich Stereocomparator is not now possible. The foregoing plate is to be understood more as a trigger for further inquiries concerning pre-Anderson positrons, than, as had until recently been felt, an actual disclosure of such a particle. For, whether justified by the event or not, the present author was encouraged by this photograph to seek further pre-1930 evidence of positive electrons. It can be confidently reported that the tell-tale tracks were 'encountered' by researchers like Orban, Rochester, Terroux, Feather, and certainly by Meitner, and the Joliot-Curies. Indeed, in pointing out the mistake in Dirac's reports and recollections concerning 1926–27, Skobeltzyn refers me to the work of Joliot and Curie of early 1932 in which mention is made of 'backward electrons'.[1]

Skobeltzyn suggests that Dirac may have confused this work of Joliot and Curie with his memory of the 1926–27 lecture at the Cavendish. Perhaps. But Blackett, Meitner, and others, certainly encountered electrons, before Anderson's discovery, which they described as 'falling back into the source', 'curving the wrong way', 'coming up from the floor', or 'moving backwards'. Skobeltzyn disagrees with my contention that several microphysicists saw, but did not observe, positron tracks prior to 1930. He writes:

None of them published their results earlier than in 1930. Besides, Williams and Terroux used in their work RaE, a source which does not emit γ-rays. Meitner–Filipp published their observation in 1933 and certainly did not use the Wilson chamber combined with a magnetic field until 1932. Ellis never worked with a Wilson chamber. Joliot-Curie began the observation with a Wilson chamber placed in a magnetic field only in 1932 (certainly not earlier than the end of 1931, i.e. definitely after my departure from Paris in August 1931). D.S. [note to the author, 22 October 1960. See Appendix IV.]

Professor Skobeltzyn's views must, of course, be given the greatest weight. Against this, however, must be balanced the reports of Dirac, Blackett and Bethe, all of whom have expressed to me their conviction that tracks were encountered, but not identified, long before Anderson's discovery. Since this chapter seeks to provide an historical and conceptual context for this reported conviction, I have placed Skobeltzyn's views alongside those of other physicists to give the reader exactly the same data that the author has.

In this connexion, the following is relevant:

I heard about 'electrons falling back into the source' at a colloquium at the Cavendish.... Someone was reporting on the work of Skobeltzyn, the Russian physicist.... I would suggest you look up *Science Abstracts*

for the years 1925–30...[Dirac, letter to the author, 12 August 1955].
...all sorts of people saw positrons well before Anderson. Blackett
certainly did and was always puzzled by the number of electrons that
came upwards from the floor...[D. H. Wilkinson, letter to the author,
12 December 1955]. ...Nobody took 'experimental' positive electrons
before 1933 seriously, nor did Dirac make the connexion in 1931. The
theoretical development was certainly taken much more seriously...
[H. Bethe, communication to the author, early 1956].

Whenever seen, such tracks were discounted as 'spurious', or as
'dirt effects'. Certainly, no experimental physicist before late 1932
made any such track his prime object of study. Part of the function
of our inquiry will be to understand why this is so, why such tracks
were always overlooked, underevaluated, or explained away. Why
did the idea of a positively charged electron seem so completely
untenable?

A

The Anderson particle

Had anyone ever seriously suggested that Professor Anderson's
discovery of the positive electron was contingent upon the prior
publication of Dirac's positron theory in 1931 [I once suggested
this], Anderson could easily refute him. Neither Dirac's book
Quantum Mechanics (1930), nor his great paper on the spinning
electron (1928)[1] had even been read by Anderson before 1932.
Indeed, he claims not to have fully understood either of these works
even when he read them later, having been encouraged to do so by
the publication of Blackett and Occhialini in 1933. Hence it could
hardly be claimed that Anderson's discovery of the positron on
2 August 1932, when he was assisted by Dr Seth Neddermeyer,
was the result of Dirac's theoretical breakthrough. No, Anderson's
discovery was of a different kind.[2]

The celebrated photograph (sketch shown in fig. 3) is that for
which Carl Anderson is honoured. But the honours accrue not
simply to his camera technique and experimental ingenuity.[3] For
here was a cloud chamber track which might also have been rejected as 'spurious' or even as a 'dirt effect'. Soon we will examine
the reasons for which Anderson's predecessors and contemporaries
were so reluctant to admit a positive electron. Theory was against
it. Observation was against it. The giants of physics, Bohr and
Rutherford, were against it.

But Anderson had the courage, and the cool logic, to be for it. Although he knew of no established theoretical route along which to infer the existence of a positron electron, Anderson's honest and rigorous interpretation of the photograph schematically reproduced above forced him to hack out his own argument *ab initio*:

...the track...cannot possibly have a mass...of a proton...[...the length...is...ten times greater than...a proton path of this curvature]...assume that at exactly the same instant...two independent electrons happened to produce two tracks...to give the impression of a single particle....This...was dismissed on a probability basis...we also discarded...the assumption of an electron of 20 million volts entering the lead on one side and coming out with...60 million volts. Other photographs were obtained which could be interpreted logically only on a positive-electron basis... (*op. cit.* p. 491).

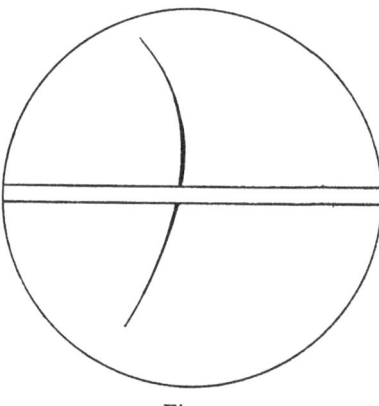

Fig. 3

In short, Anderson reasoned that the particle was coming up from below: it lost energy in the plate and curved more markedly towards the negative pole of the transverse magnetic field surrounding the chamber. This established that the particle, already identified as an electron by its range, was positively charged.

In his distinguished cosmic ray work of early 1931 Professor R. A. Millikan had already obtained two remarkable photographs, sketched below.[1] These record incoming cosmic rays encountering an atomic nucleus, out of which are projected a positive and a negative particle. Millikan's photographs (fig. 4) established that the nucleus is important in cosmic-ray absorption; moreover, both positives and negatives can be ejected from a nucleus hit by a cosmic ray. In 1919 Rutherford had shown that α-particles could dislodge protons from the nuclei of light atoms. This was the nearest analogy to Millikan's observations. But cosmic rays are not streams of α-particles. So here was a new phenomenon.

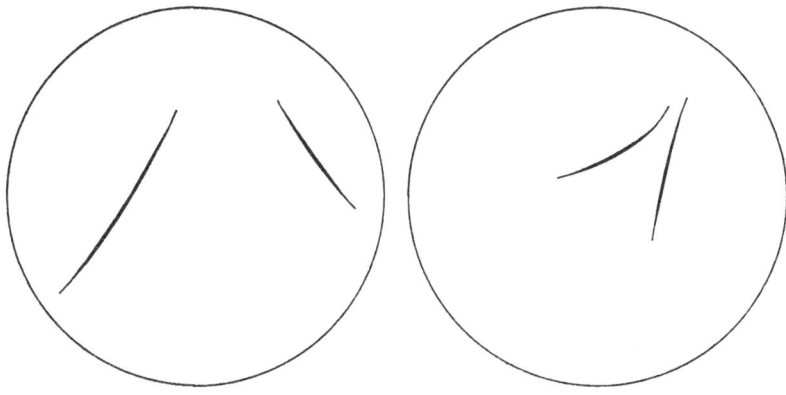

Fig. 4

In 1931, during lectures at Cambridge and Paris, Millikan interpreted the positive components of these double tracks as protons. The photographs were published on 18 December 1931.[1] This interpretation did not work with the second of the two photographs. The positive particle had little curvature; Millikan was forced to conclude that it was a proton of 450 MeV while the negative track was taken as an electron of 27 MeV. At this energy the ionization of a proton should be between 1·5 and 2 times that of an electron of 27 MeV. But there was *no detectable difference* in ionization between the positive and negative particles. This fact ran counter to certain very well established correlations.[2] Millikan thought there was something wrong with the theory of how the ionization of protons varies with energy. He did not then, nor did Anderson, ever suppose that the positive particle might be other than a proton. Millikan and Anderson agreed to come back to this 'proton' to discover the 'error'. But even now they knew that since an isolated single track of small ionizing power curved rapidly to the right, and hence ought to be a positive particle going down, the track might also be considered as having been made by a negative going up. It was necessary to have two tracks branching downward from a common centre for any positive identification of a positive as opposed to a negative particle.[1] Only a small percentage of all the particles photographed by Anderson and Millikan were so 'associated'. Therefore, the 'proton' interpretation of the second photograph was difficult to check. Moreover, for reasons to emerge, the assumption that the fundamental unit of positive electricity *could only* be associated with the proton, made active challenging of the Millikan–Anderson interpretation seem *prima facie* implausible.[2]

Then came the photograph of 2 August (cf. above). After encountering some further unambiguous cases Dr Anderson published his discovery.[3]

None the less, the appearance of 'free' positrons continued to be regarded as a rare and peculiar event, even by Anderson. Indeed, it is reliably reported that when Bohr and Rutherford heard of Anderson's publication, they were full of scepticism. Prior theoretical commitments prepared them to feel *a priori* that a wrong interpretation must have been given to whatever it was that Anderson photographed. So most high energy positive tracks encountered during subsequent months were interpreted as protons. During all this rigorous questioning and re-interpretation, the best photograph remained the original one, that of 2 August 1932.

In early 1933 Anderson published his most notable report, 'The Positive Electron'.[4] There, besides reproducing the original photograph, as and setting out his hard reasoning in favour of a positron—reasoning quoted earlier—Anderson set out also a photograph as shown in fig. 5. The outside (far left and far right) tracks are 'associated in time'. That is, they represent ionizations which certainly occurred at the same instant; they are all diffuse to the same degree.[5] The track at the extreme left is of a negative electron of 18 MeV. That at the extreme right is of a positive electron of 20 MeV. The diffuseness of both tracks results from the cloud chamber expansion having occurred $\frac{1}{2}$ sec after the passage of the ionizing particles. The ions have thus diffused away from the original track before water-vapour condensed on them. Hence, under a microscope, one can accurately count the ions per centimetre of path, as well as measure the curvature. These two considerations determined both the charges and masses of the two particles as exactly the same to within 10 and 20 per cent respectively.[1] Within these limits positive and negative electrons now seemed 'identical twins', the only difference being the sign of their charge. Indeed, in September of 1933 Anderson writes:

The striking fact that in the cosmic rays positives and negatives occur in practically equal numbers and have apparently a similar energy distribution is in accord with the assumption that the nuclear effects involved give rise to the positives and negatives in pairs (in some cases several pairs as evidenced by the showers)....[*Phys. Rev.* XLIV (1933), 416.]

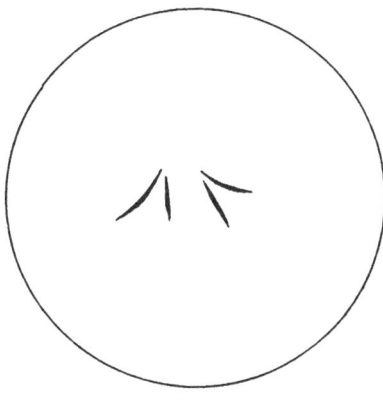

Fig. 5

This, then, constitutes part of the chronology of the discovery of the Anderson positron. Fully to appreciate the concentration, and the challenge to theoretical orthodoxy which Anderson's interpretation of his data required—this will only begin fully to come out as one explores the conceptual contexts of two further discoveries. These were the discovery of the Dirac particle, and the Blackett–Occhialini particle.

B

The Dirac Particle

The 'new' quantum mechanics of Schrödinger and Heisenberg was not in agreement with experiment during late 1927 and early 1928. When applied to the atom with point-charge electrons, it failed to account for duplexity phenomena, and the empirically discovered spin-angular-momentum ($\frac{1}{2}$ quantum).[2]

In one of the greatest theoretical papers in the history of physics, Dirac begins by saying:

One would like to find some incompleteness in the previous methods of applying quantum mechanics to the point-charge electron such that, when removed, the whole of the duplexity phenomena follows without arbitrary assumptions. . .

He goes on:

. . .the incompleteness of the previous theories [lies] in their disagreement with relativity, or alternatively, with the general transformation theory of quantum mechanics.[1]

In a most effective contribution to the new quantum mechanics, Gordon assumes that the quantum-theoretic operator of the wave equation should be got from the classical relativity Hamiltonian by putting into F, $W = ih(\delta/\delta t)$,

$$pr = -ih(\delta/\delta x_r). \quad [r = 1, 2, 3].$$

This gives

$$F\psi = [\{ih(\delta/c\delta t)+e/c(A_0)\}^2+\Sigma_r\{-ih(\delta/\delta x_r)$$
$$+e/c(A_r)\}^2+m^2c^2]\psi = 0, \quad (1)$$

ψ being a function of X_1, X_2, X_3, t.[2] Gordon's equation was valuable and influential. But it involved two difficulties which Dirac remarks with clarity. The first concerns the physical interpretation of ψ. Gordon and Klein[3] made the aforementioned assumption, and it succeeded with the emission and absorption of radiation. But it was not general enough to answer the question 'What is the probability of a dynamical variable at t having a value lying between specified limits, when the system is represented by a given wave function ψ_n?'

. . .the Gordon–Klein interpretation can answer such questions if they refer to the position of the electron. . .but not if they refer to its momentum, or angular momentum, or any other dynamical variable. We should expect the interpretation of the relativity theory to be just as general as that of a non-relativity theory. (Dirac, *ibid.* p. 612.)

The crux is this: in non-relativity quantum mechanics the wave equation is linear (of the first order in time) 'so that the wave function at any time determines the wave function at any later time. The wave equation of the relativity theory must also be linear in W if the general interpretation is to be possible' (*ibid.*). In short, Gordon's equation was not general enough: it could not give values to the dynamical variables when the system is represented by ψ_n.

The 1926 equation was of the second order in time (i.e. in $\partial/\partial t$). It is, of course, mathematically impossible to apply the general theory of transformations to differential equations higher than the first order in time.

So the first of the two difficulties Dirac perceives in the earlier Gordon equation consists in its not being of the first order in $\partial/\partial t$. He cannot be satisfied with such an equation.

The second difficulty in Gordon's interpretation arises from the fact that if one takes the conjugate imaginary of [(1)] one gets

$$[(-(W/c)+(e/c)A_0)^2+(-p+(e/c)A)^2+m^2c^2]\psi = 0,$$

which is the same as one would get if one put $-e$ for e. The wave equation (*ibid.* p. 611) thus refers equally well to an electron with charge e as to one with charge $-e$. If one considers for definiteness the limiting case of large quantum numbers, one would find that some of the solutions of the wave equation are wave packets moving in the way a particle of charge $-e$ would move on the classical theory, while others are wave packets moving in the way a particle of charge e would move classically. For this second class of solutions W has a negative value.[1] One gets over the difficulty on the classical theory by arbitrarily excluding those solutions that have a negative W. One cannot do this on the quantum theory, since in general a perturbation will cause transitions from states with W positive to states with W negative. Such a transition would appear experimentally as the electron suddenly changing its charge from $-e$ to e, a phenomenon which has not been observed. The relativity equation should thus be such that its solutions split up into two non-combining sets, referring respectively to the charge $-e$ and the charge e (*ibid.* p. 612).

In short, the otherwise effective Gordon equation, besides being of the second order in time, also generated solutions involving particles with 'negative energy'. Such electrons had not been observed, or at least not identified, by 1928. Dirac says: 'In the present paper we shall be concerned only with the removal of the first of these two difficulties.' This last remark is historically interesting. Positron transitions are often referred to in the literature of theoretical physics as 'Dirac jumps'. But, considering the occasion for Dirac's problem, and his specific attack, it would be more appropriate to call such transitions 'Gordon jumps'. The original 'negative energy solutions' were in print in 1926.[1] The problem was one of two which arose within the new quantum mechanics; it was the problem Dirac chose *not* to deal with in 1928. His problem was quite different. It was to 'obtain a wave equation of the form $(H - W)\psi = 0$ which shall be invariant under a Lorentz transformation and shall be equivalent to $[(1)]$ in the limit of large quantum numbers...' (*ibid.* p. 613).

The awe with which Dirac's 1928 paper was greeted by physicists was a function of the successful solution he found for this one problem. In an elegantly powerful demonstration, he designs a new wave equation $(i\Sigma\gamma_\mu p'_\mu + mc)\psi = 0$, which is not only of the first order; it can also be brought back into the form of the original Gordon equation, so that the important results of the latter stand out independent of the frame of reference used.[2] Dirac immediately proves the new equation sufficient to account for all duplexity phenomena; he works the spin-angular-momentum ($\frac{1}{2}$ quantum) into the logical structure of his new equation, instead of being, what it had been, a merely empirical adjustment, theoretically *ad hoc* and unconnected with general electron theory.[3] Beyond all this, Dirac's equation gave the correct results for the hydrogen atom; the Zeeman effect became calculable and explicable; the fine-structure formula comes out as a consequence; and the entire Dirac electron theory was applied (by Klein and Nischina) to the Compton scattering of electrons.

Theoretical physics has rarely witnessed such a powerful unification of concepts, data, theories, and intuitions: Newton and Universal Gravitation; Maxwell and Electrodynamics; Einstein and Special Relativity; Bohr and the hydrogen atom; these are the high spots before Dirac. From a chaos of apparently unrelated facts and ideas, Newton in his way, and now Dirac in his, built a logically powerful and conceptually beautiful physical theory.

For this reason Dirac's paper is regarded by theoretical physicists as one of 'the greatest papers of our century'.[4] However, the second difficulty within the Gordon equation, and hence the Dirac theory, remained. Dirac felt this to be a serious blemish.

In the same way, and for the same reason, Dirac sought to deal with this perplexity as Anderson and Millikan had originally tried to deal with theirs.[5] Dirac, following a suggestion of Weyl, tried to 'cook' the negative-energy solutions so that they might be construed as protons.[1] Had he been successful, his paper might have seemed even greater in 1928. For then, it would have been *the* complete theory of elementary particles; protons and electrons were then thought to be 'all there were'.

But then Weyl[2] demonstrated this interpretation to be untenable. Whatever might be the further properties of the 'negative energy' particles, they had to have the same mass as the negative electrons. This was built into the logic of the Dirac paper, and this could not be tinkered with without wrecking the symmetry of *the* entire argument. So any appeal to protons was now blocked.

At this time, Gamow, in his serio-comic manner, dubbed the new particles 'donkey electrons'. Taking them to be electronic in every other way, Gamow simply accepted that they had negative energy. Like donkeys, the harder you pushed them away, the more forcefully they came at you. Hence, a direct hit with a γ-ray photon should send the donkey electron 'recoiling' back into the source with high velocity. Thus Gamow's suggestion!

But this hypothesis, and the negative energy theory in general, was physically unintelligible. It hardly helped the experimenter who had just witnessed an electron falling back into the source to be told that it had negative energy. Moreover, the idea that two particles, one of positive energy and one of equal negative energy, could be created simultaneously out of *nothing*—this constituted a fundamental objection to the very concept of a negative energy particle.

Accordingly, Schrödinger sought to modify the Dirac equation.[3] He aimed to restructure the algebra so as to block such awkward consequences, but not in so transparently *ad hoc* a way as to constitute a mere rejection. Although at first promising, the results were shown, I think by Oppenheimer, not to be Lorentz-invariant. It is philosophically interesting to realize that this lack of relativistic invariance was, by itself, enough to mark down the Schrödinger modification as a failure. It was more important to theoreticians to have an equation which, although it harboured unintelligible solutions, was none the less Lorentz-invariant, than to adopt an alternative equation which, although easily interpretable physically, lacked invariance. The requirements of 'operationalism' alone are sometimes not quite enough to force a change in a broad-based theory.

Now the negative-energy solutions began to worry Dirac in earnest. They could not be rejected, nor written out of the theory, à la Schrödinger; the very design of an otherwise successful equation would collapse by either such attempt. Nor could they merely be christened 'donkey electrons', possessed of 'negative energy'; this solves nothing since the idea is unsound. Nor could these solutions be 'cooked' into protons, as Dirac once hoped; Weyl showed that. So these perplexing solutions could no longer be lived with comfortably.

In 1930 Oppenheimer parallels the march of Dirac's thought:

...Dirac has suggested [*Proc. Roy. Soc.* A, cxxvi (1930)] that the reasons the transitions of an electron to states of negative energy, which are predicted by his theory of the electron, do not in fact occur is that nearly all the states of negative energy are already occupied. Dirac has further shown that the unoccupied states of negative energy have many of the properties of protons...they may be represented by wave functions which would be taken to correspond to a particle of positive charge and positive mass...the mass associated with these gaps is not necessarily the same as that of the electron [N.B. Weyl showed this view to be erroneous.—N.R.H.], and he has suggested the assumption that the gaps are protons....According to Dirac the scattering takes place by a double electron jump, in which a negative electron [i.e. a negative-energy electron] jumps up into some state of positive energy and the original positive [i.e. ordinary] electron falls down into the gap left. There are several grave difficulties which arise when one tries to maintain the suggestion that the protons are the gaps of negative energy, and that there are no distinctive particles of positive charge....[1]

Oppenheimer here refers to Dirac's 'proton-cooking' phase; he also is on the verge of indicating, what Weyl was also concerned to do, that this move cannot work.[2]

If we return to the assumption of two independent elementary particles, of opposite charge and dissimilar mass, we can resolve all the difficulties raised in this note, and retain the hypothesis that the reason why no transitions to states of negative energy occur, either for electrons or for protons, is that all such states are filled. In this way, we may accept Dirac's reconciliation of the absence of these transitions with the validity of the scattering formulae (*loc. cit.*).

Slightly earlier, Oppenheimer had expressed his attitude toward the 'Dirac jumps'.

On the present theory there is no normal state for matter, because states of infinite negative energy are possible; one may in fact show that, on the present theory Dirac jumps to such states from states of positive energy—jumps in which the energy and momentum lost by the matter are taken up by the field—are not only possible, but infinitely probable. But that the theory should have predicted this is a token of an error in the theory....[1]

It is to this 'error in the theory' that Oppenheimer, in the previous quotation, addressed himself.

The Dirac conjecture responsible for Oppenheimer's reactions can now be set out:

The most stable states for an electron (i.e. the states of lowest energy) are those with negative energy and very high velocity. All the electrons in the world will tend to fall into these states with emission of radiation. The Pauli exclusion principle, however, will come into play and prevent more than one electron going into any one state. Let us assume that there are so many electrons in the world that all the most stable states are occupied, or more accurately, that all the states of negative energy are occupied except perhaps a few of small velocity. Any electrons with positive energy will now have very little chance of jumping into negative-energy states, and will therefore behave like electrons are observed to behave in the laboratory. We shall have an infinite number of electrons in negative-energy states and indeed an infinite number per unit volume all over the world, but if their distribution is exactly uniform we should expect them to be completely unobservable. Only these small departures from exact uniformity, brought about by some of the negative-energy states being unoccupied, can we hope to observe.[2]

Note Dirac's final remark. Here, in January of 1930, he is entertaining what would be necessary in order to make observations of 'negative energy states' which he now felt obliged to accept, since his electron theory refused to settle for less. It is the outside possibility of such observations that Dirac is now beginning to take seriously, even though the 'negative energy' idea is still a worry.

In 1930 and 1931 Oppenheimer and Heisenberg (while writing on Halogen-theory) began to speak of these small departures from exact uniformity as 'vacancies in the negative-charge continuum'. These thinkers, along with Dirac, began to think of the $+e$ (or $-W$) solutions of 1928 not so much as negative energy solutions, but simply as positive charge solutions. By early 1933 Blackett was confidently referring to the 'vacancies' as 'holes'.[3] These 'holes', gaps in an infinite sea of negative electricity, were taken now to be ordinary particles with positive kinetic energy and positive charge. It was already understood that they had the mass of an electron. The conclusion, therefore, was inescapable; the 'negative-energy solutions' heralded nothing other than electrons with a positive electric charge. This shift of ideas percolated only slowly into the strange seas of physical thought in 1931. Thus Halpern and Thirring write:

Dirac's system of equations refers to particles of charge $+e$ as well as to those of charge $-e$; but this means that it has too many solutions for the charge $+e$. As neither $+e$ nor $-e$ has an advantage over the other, exactly half of the solutions refer to a positively charged particle and the other half to a negatively charged particle. Since in general two different solutions may be combined together, that is, a particle can pass from a state characterized by a proper function belonging to $+e$ to the state characterized by a proper function belonging to $-e$, this signifies that according to Dirac's theory the electrons can change their sign. This is an effect for which no experimental evidence has ever been observed. It can also be shown that in the same process, namely, in the transition of the charge $+e$ to $-e$, the sign of the total energy E must change.... Dirac's theory leads to the quite unintelligible result that particles with *negative mass* must exist....

In practice, therefore, until these difficulties are removed we shall merely retain one-half of the solutions which correspond to positive mass and subject them to comparison with experiment. That this comparison comes out satisfactorily in every respect was mentioned in the introduction....[1]

Bolder is Oppenheimer's remark of 1930: '...according to Dirac not all of the states of negative energy are full; there are a few gaps in the distribution for negative electrons merely at rest; and thus transitions to states of negative energy should not be quite excluded.'[2]

Even as late as 1933, Pauli argued that the Dirac equation was a wonderful contribution to microphysics; however, it entailed the existence of positive electrons; hence, the equation must be erroneous.[3]

Thus, the 'negative-energy' solutions, plus the rejection of their interpretation as protons, plus the 1930–32 'hole-theory', of Dirac, Heisenberg and Oppenheimer—all this constituted the gradual development of a *prediction* of the existence of positive electrons before they were observed. The context remained full of uncertainties and incalculables. None the less, the strength of

Dirac's electron theory, plus the logic of its consequences, forced theoreticians to opt for the as-yet-unobserved entity.

This development is what inclines theoreticians to characterize the discovery as a theoretical break-through. The reader is left to decide whether the final guerdon should be given to the theoretical prediction made in the total absence of a confirming observation, or to the frank interpretation of an awkward observation in the total absence of a theory to explain it. Doubtless either undertaking required uncommon courage, skill, conviction and care. Doubtless also, both Carl Anderson and Paul Dirac possessed these characteristics to a remarkable degree. This is shown by the fact that, in entire independence of each other, and against the best established reasons to the contrary, they took their stand on the existence of a positively charged electron. Let us conclude this section with a quotation from Dirac himself, as he writes (in 1931) of the steps leading up to his great prediction:

The mathematical formalism at that time involved a serious difficulty through its prediction of negative kinetic energy values for an electron. It was proposed to get over this difficulty, making use of Pauli's Exclusion Principle which does not allow more than one electron in any state, by saying that in the physical world almost all the negative-energy states are already occupied, so that our ordinary electrons of positive energy cannot fall into them. The question then arises as to the physical interpretation of the negative-energy states, which on this view really exist. We should expect the uniformly filled distribution of negative-energy states to be completely unobservable to us, but an unoccupied one of these states, being something exceptional, should make its presence felt as a kind of hole. It was shown that one of these holes would appear to us as a particle with a positive energy and a positive charge and it was suggested that this particle should be identified with a proton. Subsequent investigations, however, have shown that this particle necessarily has the same mass as an electron[1] and also that, if it collides with an electron, the two will have a chance of annihilating one another much too great to be consistent with the known stability of matter.[2]

It thus appears that we must abandon the identification of the holes with protons and must find some other interpretation for them. Following Oppenheimer,[3] we can assume that in the world as we know it, all, and not merely nearly all, of the negative-energy states for electrons are occupied. A hole, if there were one, would be a new kind of particle, unknown to experimental physics, having the same mass and opposite charge to an electron. We may call such a particle an anti-electron. We should not expect to find any of them in nature, on account of their rapid rate of recombination with electrons, but if they could be produced experimentally in high vacuum they would be quite stable and amenable to observation. An encounter between two hard γ rays (of energy at least half a million volts) could lead to the creation simultaneously of an electron and anti-electron, the probability of occurrence of this process being of the same order of magnitude as that of the collision of the two γ rays on the assumption that they are spheres of the same size as classical electrons. This probability is negligible, however, with the intensities of γ rays at present available.

The protons on the above view are quite unconnected with electrons. Presumably the protons will have their own negative-energy states, all of which normally are occupied, an unoccupied one appearing as an anti-proton.

In this passage Dirac sharply distinguishes the proton from the positron. He states that the latter, as well as the former, is in principle amenable to observation, and predicts circumstances under which positrons might be detected. He entertains seriously the 'materialization' hypothesis later discussed by Joliot-Curie: this is nothing less than the creation of matter out of energy, one of the boldest conjectures of all time. And he considers the possible existence of an anti-proton. All this in less than forty lines of prose!

C

The universal reluctance to accept a new particle

How is one to explain the uniform resistance everywhere shown to the idea of a 'third' particle? Or, as Millikan puts it: 'Prior to the night of 2 August, 1932, the fundamental building-stones of the physical world had been universally supposed to be simply protons and negative-electrons. Out of these two primordial entities all of the 92 elements had been formed.'[1]

How is one to explain the fact that Rutherford, the experimentalist, and Bohr, the theoretician, felt that whatever Anderson actually observed its interpretation as a new particle must be in error?

How is one to explain Dirac's energetic attempts to construe the 'negative-energy' solutions as representing protons? His ultimate concession of a new particle was diffident and reluctant. Schrödinger too, did everything to avoid this 'third particle' conclusion.

And recall Oppenheimer's wish to ' . . . return to the assumption of two independent elementary particles, of opposite charge and dissimilar mass. . . . '

And how is one to explain the pre-Anderson photographs of tracks now readily identifiable as having been made by positive electrons? Most of these remained unnoticed, or were characterized as 'spurious', or 'dirt effects'; or they were left uninterpreted, or were wrongly interpreted, as with some tracks seen by Skobeltzyn, Blackett and the Joliot-Curies.

There must be some general explanation of this reaction against there being a positive electron. Explore with me, on a slight detour from our main route, the conceptual background which was designed to exclude a third particle. Perceiving this will throw the discoveries of Anderson, Dirac and Blackett into high relief. It will also explain the quickening of research and experimentation within particle theory during the 1930's. It will, furthermore, lace together our earlier chapters with the present one.

Our answers lie partly within the development of the theory of electricity.

'Electron' was the ancient Greek word for amber. The Greeks knew that rubbed amber attracts other objects to it. Gilbert (1600) found that glass, when rubbed, had the same property. The glass was 'electrified', that is, amberized. Dufay (1698–1739) discovered that sealing wax, rubbed with cat's fur, was also electrified, but differently from the glass. The wax attracted all electrified bodies which had been repelled by the glass, and repelled those attracted by the glass. Dufay remarks that here are ' . . . two electricities of a totally different nature . . . , transparent solids . . . , or resinous bodies: . . . Each of them repels bodies which have contracted an electricity of the same nature as its own, and attracts those whose electricity is of the contrary nature.' Dufay called these two types of electricity 'resinous' and 'vitreous'—names still used. In 1747 Benjamin Franklin also recognized these two forms of electricity. He arbitrarily called them 'positive' and 'negative'. A body was then negatively charged if repelled by sealing wax rubbed with cat's fur. It was positively charged if repelled by a glass rod rubbed with silk. These definitions still obtain.

Franklin also discovered the conservation of electric charge. He noted that electricity is not created by rubbing the glass, but only transferred to the glass from the material rubbed against it. The latter loses what the glass gains. Hence the total quantity of electricity in any insulated system is invariable. Franklin considers the material rubbed against the glass as deficient in electricity; he says it has a charge of $-e$. But the glass, however, now has a superfluity of electricity, and a charge of $+e$. So Dufay's 'vitreous electricity' is identical with Franklin's electrical fluid. And 'resinous electricity' is the absence of that electrical fluid, or rather, a deficiency of that 'fluid' possessed by all ponderable bodies. Thus Franklin's electrical fluid exists as a natural constituent of all matter in the neutral, unelectrified state. A superfluity of this naturally appearing electrical fluid gives the body a positive charge. A deficiency results in negative charge.

Here, then, is a clear conception of electrification, of positive and negative charge.

The possibility of electricity also being particulate, or granular, or atomic, in character gets serious attention in the nineteenth century. In 1851 Joule determined the average speed of molecules moving in a given gas at ordinary temperature.[1] In 1860 Clerk Maxwell determined the mean distance a molecule moves between collisions.[2] Clausius had already discussed this, but Maxwell gave the first evaluation, utilizable within the general theory of the viscosity of gases. He also estimated the number of molecules in a cubic centimetre of gas. Ostwald resisted every form of the atomic hypothesis until the last edition of his *Outlines of Chemistry*. The isolation and counting of gas ions, the Brownian motion, and the kinetic theory ultimately forced him to accept the particulate position.

Even Franklin had written 'the electrical matter consists of particles extremely subtle, since it permeates common matter, even the densest, with such freedom and ease as not to receive any appreciable resistance'. In 1833 Faraday found that a given amount of electricity passing through a solution containing hydrogen, or a compound of hydrogen, caused the same amount of hydrogen to appear at the negative terminal, irrespective of the compound or the strength of the solution. Moreover, the quantity of electricity required to make 1 g of hydrogen appear always deposited from a silver solution exactly 107·05 g of silver. Since the silver atom is 107·05 times heavier than the hydrogen atom, and both are associated in the solution with exactly the same amount of electricity, it was concluded that all univalent atoms, i.e. those which combine with one atom of hydrogen, carry the same amount of electricity. All bivalent atoms carry twice this amount. Valency, then, is proportional to the amount of electricity carried by the atom in question.

However, all the scientific reflexion up to this time took electricity to exist *on* a charged body. It was assumed to exert forces on other charged bodies just as does gravity. To Faraday this was nothing but action at a distance; it was totally repugnant to him. He learned that the electrical force between two charged bodies depends in some way on the nature of the intervening medium. Gravitational pulls are independent of the medium. Consequently, electrical forces are transmitted rather as the elastic deformation at one end of a rod is transmitted along the rod.

Clerk Maxwell puts the point thus: '...it is in questionable scientific taste, after using atoms so freely to get rid of forces acting at sensible distances, to make the whole function of the atoms an action at insensible distances...'.[1]

However, unlike transmission of elastic deformations, electrical forces act through a vacuum. This led directly to the infamous ether, whose history it is not here our purpose to discuss.

These concepts were put into a powerful mathematical form by James Clerk Maxwell. Attention was now drawn from phenomena in or on a conductor and refocused on the 'stresses and strains' in the medium surrounding the conductor. When Hertz, in 1887, found that electrical forces were transmitted as waves, at precisely the speed of light predicted by the Faraday–Maxwell theory, the triumph of the ether-stress view seemed complete. Professor Oliver Lodge, indeed, says that electrical charge is only 'a state of strain in the ether', and a current is nothing other than a 'continuous breakdown of a strain in the medium within the wire'.[2] But this 'strain theory' led men to think of the strain distributed evenly over the surface of a charged body, rather than as it was observed to do, namely, radiate from centres all over the surface of the body.

Concerning the particulate character of electricity, then, between 1833 and 1900 physicists:

(1) Thought of the passage of electricity through a solution as a

motion of specks or atoms, each carrying an exact multiple, between one and eight, of some elementary electrical atom.

(2) Thought of the passage of a current through a metal as a continuous 'slip' or 'breakdown' of a strain in the matter itself.

(1) and (2) were felt to be different in kind; electrolytic conduction and metallic conduction. The latter was much more prominent. Hence the atomic conception was only slightly heeded. Maxwell notes the problem: He says, 'for convenience in description we may call this constant molecular charge (revealed by Faraday's experiments) one molecule of electricity'.[1] But he then goes on to deny any real physical significance to this convenient decision, saying 'it is extremely improbable that when we come to understand the true nature of electrolysis we shall retain in any form the theory of molecular charges, for then we shall have obtained a secure basis on which to form a true theory of electric *currents* and so become independent of these provisional hypotheses.'

And, after all, Maxwell was justified in granting that an ion always took into the solution a definite quantity of electricity, without thereby conceding that the charge on the electrode is made up of the same number of electrical atoms.

At about the same time, Weber[2] built up an entire theory on the modified Franklin idea. Weber explained all electrical phenomena of conduction, including thermodynamic and Peltier effects, by assuming two types of electrical constituents of atoms. One of these was taken to be more mobile than the other. Thus Ampère's molecular current, which rendered molecules into tiny electromagnets, Weber conceives as the rotation of light positive charges about heavy negative ones.

He writes:

The relation of the two particles, as regards their motions, is determined by the ratio of their masses e and e', on the assumption that in e and e' are included the masses of the ponderable atoms which are attached to the electrical atoms. Let e be the positive electrical particle. Let the negative be exactly equal and opposite, and therefore denoted by $-e$ (instead of e'). But let a ponderable atom be attracted to the latter so that its mass is thereby so greatly increased as to make the mass of the positive particle vanishingly small in comparison. The particle $-e$ may then be thought of as at rest, and the particle $+e$ as in motion about the particle $-e$. The two unlike particles in the condition described constitute then an Amperian molecular current. (Weber, *loc. cit.*)

Except for the fact that it is the *negative* particle whose mass is negligible compared with the positive, and not vice versa as Weber conjectures, the quotation above contains precisely the idea which Lorentz later developed into modern electron theory.

In 1874 Stoney stated the atomic theory of electricity. He even estimated the value of the elementary electrical charge [$0 \cdot 3 \times 10^{-10}$ absolute electrostatic units].[1] Stoney writes:

...nature presents us with a single definite quantity of electricity which is independent of the particular bodies acted on....I shall express Faraday's law in the following terms...: *for each chemical bond which is ruptured within an electrolyte a second quantity of electricity traverses the electrolyte which is the same in all cases.* This definite quantity of electricity I shall call e_1. If we make this our unit of electricity, we shall probably have made a very important step in our study of molecular phenomena.... (*loc. cit.*)

And compare Helmholtz writing in 1881: '...if we accept the hypothesis that the elementary substances are composed of atoms, we cannot avoid concluding that electricity also, positive as well as negative, is divided into definite elementary portions which behave like atoms of electricity.'[2]

Even Kelvin, otherwise disposed to regard electricity as a 'continuous homogeneous liquid', writes: 'Faraday's laws of electrolysis seem to necessitate something atomic in electricity....'[3]

The term 'electron' was coined in 1891 by G. Johnstone Stoney. This was his designation for the 'natural unit of electricity', that is, that quantity of electricity which must pass through a solution in order to liberate at an electrode one atom of hydrogen or of any univalent substance. 'A charge of this amount is associated in the chemical atom with each bond. There may, accordingly, be several such charges in one chemical atom, and there appear to be at least two in each atom. These charges which it will be convenient to call "electrons" cannot be removed from the atom, but they become disguised when atoms chemically unite.'[4]

Consider the semantical moral behind this short history. For Stoney 'electron' denoted the elementary quantity of electricity; this and nothing more. Hence Stoney feels that every atom contains at least two such electrons, one positive and one negative. Otherwise ordinary matter would not remain electrically neutral. But 'electron' was rapidly appropriated for other uses as well. The dynamical particle we now call the 'free negative electron', has a mass of 1/1835 of the hydrogen atom. This massy particle also 'has' the fundamental unit of electricity; $e = 4 \cdot 80 \times 10^{10}$ e.s.u. That is, the particulate electron also *has* the electron as one of its properties! Only this latter property was intended by Stoney as the electron. The 'physical object' connotations so prominent in the work of J. J. Thomson were just becoming apparent at the turn of the century.

However, Stoney's use of 'electron' continued in use. Thus Thomson, Rutherford, Campbell and Richardson, particularly after 1913, always talk of positive as well as of negative electrons. The mass associated with the former, that is, the positive electron, was usually that of the hydrogen atom itself. But these physicists are always referring to the positive elementary charge (found on the nucleus of the hydrogen atom), and the negative elementary charge (found on the particle called 'electron'). Nernst clearly defined the positive and negative electrons as the elementary positive and elementary negative electrical charges respectively.[1]

So, during the First World War, 'electron' could be construed either as the fundamental unit of electrical charge (positive or negative), or as the dynamical material particle one of whose properties was to carry the negative unit of that charge. Positive electrons and negative electrons were freely discussed. But by 1919, when the proton's existence had been definitely established, the terminology within elementary particle physics had become intricate. The 'positive electron' was now the proton. And the 'negative electron' was now the electron.[2]

Beyond this, however, such an intimate association between the two elementary particles, and the two basic units of electricity— this intimacy made the very idea of a particle other than the proton or the electron difficult to conceive. What could such a particle be like? To be fundamental, it too would have to possess *a basic unit of charge*. But the basic units were already possessed by the proton and the electron. Hence, the idea of a third particle just was the idea of a fundamental material body possessing a basic unit of some third kind of electricity—not positive, not negative, and not neutral, since electrical neutrality was the result of balanced combinations of positive and negative charge. Just as ' + ' and ' − ' seemed to exhaust the totality of electrical charge, so 'positive electron' and

THE POSITRON

'negative electron' seemed to do the same. Since the proton and the electron came to be thought of not only as carrying the charge, but virtually as being the charge,[1] the very conception of a third particle beyond the proton and the electron seemed insupportable.

It is this profound conceptual resistance, built into the structures of classical electrodynamics and elementary particle theory, which must be appreciated in order to understand why physicists like Dirac, Blackett, Skobeltzyn, Pauli, Oppenheimer, Anderson, Bohr and Rutherford struggled so hard to avoid having to make such a supposition. Given this, the positive electron hypothesis had to be repugnant to the physicists of 1930. All the more, then, must we admire the conceptual boldness of Anderson and Dirac who, in their different ways, at first tumbled to the conclusion that positive electrons exist, and then had the courage to stand up and be counted amidst a potentially unreceptive scientific audience.[2]

D

The Blackett–Occhialini particle

In a remarkable paper entitled 'Some photographs of the tracks of penetrating radiation',[3] P. M. S. Blackett and G. P. S. Occhialini swept away all residual resistance to the existence of positrons.[4]

They discovered that the 'Anderson particle' and the 'Dirac particle' were the *same* particle. This is a meta-physical discovery which, in the history of science, ranks high within a select class.

Blackett and Occhialini employed a new experimental method. Whereas Skobeltzyn succeeded in photographing one track in every ten expansions, and Anderson but one in fifty, Blackett and Occhialini arranged to have high speed particles take their own photographs.[5] This new technique succeeded in photographing 80 per cent of all the high-speed particles moving through the chamber. This matter of frequency was important to the question already felt so keenly by Anderson and Dirac: 'Are these the tracks of protons?'

Blackett and Occhialini sandwiched the cloud chamber between two Geiger–Müller counters. Any ray passing through both counters must also pass through the illuminated part of the chamber (*ibid.* p. 717). The time required from the discharge of the counters to the end of the expansion was but $\frac{1}{100}$ of a second.

Hence the ions diffuse only a short distance from the particle's path. The tracks were only 0·8 mm thick; this allowed accurate measurements to be made. $\frac{1}{100}$ second after expansion an illuminating flash begins; it lasts $\frac{1}{30}$ second. Seven hundred photographs revealed 500 high energy particles. Seventy-five per cent of all the photographs disclosed a single particle-track traversing both counters. Within a magnetic field of 2000 Gauss these particles remained undeflected. So their main energy was greater than 300 MeV. Other photographs, however, showed multiple tracks, such as had been originally detected by Skobeltzyn.[1] The Blackett and Occhialini photographs reveal an astonishing complexity amongst these secondary tracks. Their interpretation was difficult.

The most notable feature common to many of these multiple tracks is the occurrence of a group of several tracks diverging, mainly downwards, from some region in the material surrounding the chamber.... Sometimes a group of these tracks appear to diverge fairly accurately from a single point....When such a *shower* of particles is seen to have entered the top of...the chamber, it is not infrequently found that a subsidiary radiant point occurs in the metal plate...across the chamber. ...it is necessary to come to the same remarkable conclusion that has already been drawn by Anderson (*Science*, LXXVI (1932), 238) from similar photographs. This is that some of the tracks must be due to particles with a positive charge, but whose mass is much less than that of a proton (*ibid.* p. 703).

The most important measurement in this connexion was for Blackett and Occhialini, just as it had been for Anderson, the determination of the track curvature by a magnetic field ($H\rho$). Also, the ionization density was relevant, but much more difficult to determine. This depends only on the charge and the velocity, not on the mass. But the velocity of a particle of a given $H\rho$ depends on the particle's mass. Consequently, observing a particle's $H\rho$ and its ionization allows an accurate determination of its mass, even though the upper limit of the mass cannot be established really reliably for any given $H\rho$ range. On the basis of these considerations Anderson and Blackett reached the same conclusion concerning the new positively charged particle; it had to have a mass 'much less than that of a proton'.

The 'shower' picture depicted in fig. 6 suggests that all the particles came from the same direction; that any track moving vertically is probably proceeding downward; any particle deflected to the left is negative and any particle deflected to the right is positive. Had these particles been protons their ranges could only be between 0·2 and 3 cm in air—but they actually travel 12 cm! A study of the ionization density (using Bethe's calculations)[1] confirms this. These particles must be fast electrons; if they were protons their ionization would be from 10 to 100 times heavier than is actually observed. 'The only possible conclusion of the argument both from the range and from the ionization is that these tracks are due to positively charged particles with a mass comparable with that of an electron rather than with that of a proton' (*ibid.* p. 707).

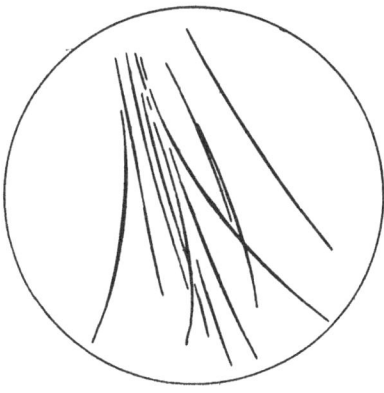

Fig. 6

Blackett and Occhialini found 14 tracks which were identified as positive electrons. Others were less certain. The ratios of the positive and negative particles were about equal in the shower photographs. In similar research Kunze had interpreted all these positively curved tracks as protons. Against this Blackett argues '...all the single tracks which do show marked positive curvature have nearly the same specific ionization as the unaffected ones, so are either negative electrons going up, or positive electrons coming down'.[2]

Of signal interest are the following remarks of Blackett and Occhialini:

...at least 7 positive and negative electrons have been found diverging from a single point in glass, copper, and lead, and presumably therefore from single nuclei...there are three possible hypotheses...about the origin of these particles: (a) they may have existed previously in the struck nucleus, or (b) they may have existed in the independent particle, or (c) they may have been created during the process of collision...it is reasonable to adopt the last hypothesis (*ibid*. p. 712).

One cannot overestimate the importance of this last conjecture. Not only do these two physicists surely identify positron-negatron pairs,[1] they also give a correct interpretation of them, and even develop a first approximation to what ultimately became (in the works of Oppenheimer, Plesset, Fermi and Uhlenbeck, 1933) the theory of *pair creation*. It is certain that at about this same time the Joliot-Curies, with their hypothesis of 'materialization', were encountering similar phenomena. And in the history of physics this is the first time that the hypothesis of creating matter out of energy was seriously entertained, now by several independent investigators simultaneously. This physical discovery in itself sets the positron history apart, and makes it wholly different from the neutrino and neutron discoveries. Blackett continues: 'If, however, the conservation of the electric charge is to be fulfilled then positive and negative electrons must be produced in equal numbers ...in this way one can imagine that negative and positive electrons may be born in pairs during the disintegration of light nuclei...' (*ibid*. p. 713).

It is at this juncture that Blackett makes his 'meta-physical' discovery, the importance of which was tremendous for subsequent elementary particle theory:

The existence of positive electrons in these showers raises immediately the question of why they have hitherto eluded observation. It is clear that they can have only a limited life as free particles since they do not appear to be associated with matter under normal conditions.

It is conceivable that they can enter into combination with other elementary particles to form stable nuclei and so cease to be free, but it seems more likely that they disappear by reacting with a negative electron to form two or more quanta.

This latter mechanism is given immediately by Dirac's theory of electrons (*Proc. Roy. Soc.* A, cxxvi (1930), 360; cxxxiii (1931), f. 60). In this theory all but a few of the quantum states of negative kinetic energy, which had previously defied physical interpretation, are taken to be filled with negative electrons. The few states which are unoccupied behave like ordinary particles with positive kinetic energy and with a positive charge. Dirac originally wished to identify these 'holes' with protons, but this had to be abandoned when it was found that the holes necessarily have the same mass as negative electrons. It will be a task of immediate importance to determine experimentally the mass of the positive electrons by accurate measurements of their ionization and $H\rho$. At present it is only possible to say that no difference between the ionization from the tracks of negative and positive electrons of the same $H\rho$ has been detected so that provisionally their masses may be taken as equal.

On Dirac's theory the positive electrons should only have a short life, since it is easy for a negative electron to jump down into an unoccupied state, so filling up a hole and leading to the simultaneous annihilation of a positive and negative electron, the energy being radiated as two quanta. ...When the behaviour of the positive electrons has been investigated in more detail, it will be possible to test these predictions of Dirac's theory. There appears to be no evidence as yet against its validity, and in its favour is the fact that it predicts a time of life for the positive electron that is long enough for it to be observed in the cloud chamber but short enough to explain why it had not been discovered by other methods.

It should be possible to find evidence on the photographs of positive electrons which have entered a metal plate but which do not emerge again owing to their annihilation while traversing the metal. It is also possible that the gamma-ray annihilation spectrum may be detectable by observations of the Compton recoil electrons. According to Dirac's theory, this spectrum should have a lower limit at an energy of 0.5×10^6 volts and should extend through a maximum at a slightly greater energy to a steadily decreasing intensity for high energies....

Again the hypothesis of the existence of positive electrons amongst the secondary particles produced by neutrons, would provide an explanation of the curious fact discovered by Curie and Joliot (*Exposé de Physique Théoretique*, 1933, p. 21) that fast electron tracks are found with a curvature indicating a negative electron moving towards the neutron source.

Here is the perfect marriage of experiment and theory. Here we have observations of positrons *with* understanding. It would be too easy, and too inaccurate, to suggest that Blackett had re-discovered, or simply 'confirmed'[1] Anderson's own work. There is something unique and unprecedented in the Blackett paper. Not only is it a discovery of much that was unknown, it is also a dis-covery of the fact that some things that were already known are interrelated in unsuspected ways. Of course, Blackett's technique allowed him to observe pair production much more completely than Anderson. The step from Blackett's observations to Dirac's insights was thus much shorter here than it had been for Anderson. Moreover, Dirac's work, at least in principle, was known to Blackett: Anderson was not in the same position to the same degree. Blackett's paper had a galvanic effect on particle theory. For one thing, as he so candidly testifies himself, Professor Carl Anderson realized that there was a connexion between his work and the research of Dirac only *after* reading the paper of Blackett and Occhialini.[1] This must have been true of many others too. For another thing, many 'spurious' phenomena now seemed amenable to possible re-interpretation.[2]

Immediately searches began for positron tracks from other sources. In April 1933, Chadwick, Blackett and Occhialini, Curie and Joliot, and Meitner and Phillipp[3] all reported that the bombardment of beryllium by polonium α rays produces radiation which results in positron production.

The theoretical work of Fermi and Uhlenbeck (1933) followed directly; the Dirac theory was thus extended so as to give a full understanding of pair creation and annihilation.[4] Moreover, once this important conceptual break-through had been achieved, once it was seen to make sense to entertain the possible existence of particles other than the proton and the negatron, elementary particle theory really began to accelerate. Now the neutron, discovered at about the same time Anderson made his discovery,[5] found its first comfortable place within the framework of particle theory. The Yukawa particle was announced in 1935.[6] In 1937 Anderson, again, discovered a particle related to that of Yukawa: these were both what are now called *mesons*. The ground was prepared for the active prosecution of meson theory. Everything in microphysics since 1937 is thus genetically connected, intimately, with the positron discovery of Dirac, Anderson and Blackett. The importance to all physics of the idea of matter-creation has already been stressed.

Some incidental morals fall gently from this long narrative. It is interesting to note that, during the period 1931–32, Dirac and Blackett were working within half a mile of each other. St John's College, Cambridge, and the Cavendish Laboratory are but a sparrow's-flight apart. Yet the actual conversation to have passed between these physicists must be adjudged very slight. Perhaps something of the same nature obtained, at the same time, at the

Norman Bridge Laboratory. For here, at Cal. Tech., we find not only Anderson and Millikan. Oppenheimer was also at Norman Bridge and later came down from Berkeley. He was intimately familiar with the Dirac theory and must also have known of the exciting tracks being produced within Anderson's machine. Just as the 'Gordon jumps' and the 'spurious tracks' existed side-by-side without being related, so this seemed to be a pattern within positron inquiry until 1933.

Of course, there may also be a sociological dimension to the conceptual history set out here. American theoretical physics was somewhat weak during this entire period. Dirac, and the Germans, would probably not have undertaken to write to Oppenheimer on their own initiative, and almost certainly not to Anderson. Perhaps such factors affect the growth of science as strongly as does any strictly conceptual or theoretical development, such as we have studied here.

A practical consequence of the positron discovery consists in the fact that, since the hypothesis of 'anti-matter' had now seemed fully confirmed, the hunt for the anti-proton[1] and anti-neutron immediately appeared feasible. In fact, the great synchro-cyclotron at Berkeley was specifically 'built up' to 6 BeV to determine whether or not the anti-matter theory, implicit in Dirac's original paper, could be confirmed for anti-protons and anti-neutrons.

One general conclusion of our immediate inquiry is this:

Those accounts of microphysical discovery which suggest them to be 'single-line' disclosures of facts, reminiscent of the naturalist finding a new bug beneath a rock—such accounts are to be held in suspicion. They distort and dwarf the enterprise of the physicist. But the conceptual complexity within physics has shaped the recent history of human thought. Our accounts of that research must respect that complexity.

APPENDIX I

From Newton's *Opticks*, 2nd edition (London, 1718):

Query 17. If a Stone be thrown into stagnating Water, the Waves excited thereby continue some time to arise in the place where the Stone fell into the Water, and are propagated from thence in concentrick Circles upon the Surface of the Water to great distances. And the Vibrations or Tremors excited in the Air by percussion, continue a little time to move from the place of percussion in concentrick Spheres to great distances. And in like manner, when a Ray of Light falls upon the Surface of any pellucid Body, and is there refracted or reflected: may not Waves of Vibrations, or Tremors, be thereby excited in the refracting or reflecting Medium at the point of Incidence, and continue to arise there, and to be propagated from thence: as long as they continue to do so,[1] when they are excited in the bottom of the Eye by the Pressure or Motion of the Finger, or by the Light which comes from the Coal of Fire in the Experiments above mention'd? And are not these Vibrations propagated from the point of Incidence to great distances? And do they not overtake the Rays of Light, and by overtaking them successively, do they not put them into the Fits of easy Reflexion and easy Transmission described above? For if the Rays endeavour to recede from the densest part of the Vibration, they may be alternately accelerated and retarded by the Vibrations overtaking them. [Pages 322, 323.]

APPENDIX II

The important feature of our electron-positron theory is that it is a quantum-field theory. This means that the fields $\psi(x)$ are operators, rather than functions representing the state of the electron. For example, $\psi(x)$ creates a positron or destroys an electron and $\psi^+(x)$ does the reverse. It is not feasible here to describe this theory in detail, but the important physical fact is that it allows electrons and positrons to be created and destroyed. A high energy γ ray ($E > 2mc^2$) passing through the Coulomb field of a nucleus can create an electron-positron pair. Such a situation cannot be described by an ordinary wave function; rather it is described by a state vector in an abstract (non-separable) Hilbert space in which the basic states have different numbers of particles. Since the Hamiltonian of the system has matrix elements between these states of different particle number, transitions such as $\gamma \to e^+ + e^-$ are possible.

Now for a few words about the non-relativistic limit where particle number is conserved and the ordinary wave function is a useful, though approximate, concept:

If the energies involved in a given physical situation are small compared to an electron's rest energy, one encounters this limit: it is the realm of most of atomic physics and chemistry (aside from small relativistic corrections which show up in fine structure, the Lamb shift, etc.). It is ordinary wave mechanics with wave functions which obey the Pauli exlusion principle—odd under interchange of particle variables. In this realm, certain classical 'primary' questions are simply answered: the electron is intrinsically a point particle with spin one-half and a definite mass. Here 'point' means that it is described by a wave function $u(\vec{x}, t)$ in which there is no variable for internal structure (other than spin). Intrinsically the electron's charge is located at a point where it has an infinite density δ (function). 'Age of an electron' has no meaning; 'solidity' has no meaning (can a point be compressed?); 'shape' has no meaning (does a point have a shape?). Of course $u(\vec{x}, t)$ gives a probability distribution for such quantities as position, momentum,

THE CONCEPT OF THE POSITRON

energy. These results are well confirmed by experiment (within its proper domain of applicability).

Perhaps an illustration will be helpful.

$P(\vec{x}, t) = |u(\vec{x}, t)|^2$ is probability distribution for $\vec{x}$; $\rho(\vec{x}, t)$ is observable charge distribution. The theory says:

$$\rho(\vec{x}, t) = eP(\vec{x}, t) = \iiint \delta(\vec{x} - \vec{x'}) P(\vec{x'}, t) d^3x'.$$

The $\delta(\vec{x} - \vec{x'})$ represents the intrinsic point charge nature of the electron. Conceivably, a different connexion could exist between ρ and P, with $\delta(\vec{x} - \vec{x'})$ replaced by $f(\vec{x} - \vec{x'})$, for example, with f a 'spread-out' δ function. Such an f would represent the intrinsic (or, if you like, internal) charge distribution of the electron. The remarkable agreement of atomic spectra with theoretical predictions of wave mechanics indicates that f is a δ function, or at least has a *very* small spatial extension. (The Lamb shift, can, in some sense, be interpreted as due to a finite size of the electron.)

Let us turn to the more complete theory. It should be emphasized again that the word 'electron' is ambiguous; that is, even now the precise definition of what we mean by 'an electron' or 'a positron' cannot be given without taking into account the interactions of the particle with other fields, such as the electromagnetic field. To illustrate, we might try to write

$$\psi = \psi_{\text{ce+}} + \psi_{\text{de-}},$$

where $\psi_{\text{ce+}}$ creates positrons and $\psi_{\text{de-}}$ destroys electrons. It is easy to verify that this separation is not independent of the interaction. For instance, if there is no interaction, one obtains a separation $\psi'_{\text{ce+}} + \psi'_{\text{de-}}$ (free representation). But if one has a Coulomb potential there is a different separation $\psi''_{\text{ce+}} + \psi''_{\text{de-}}$ (bound representation). If we try to express the state of a bound electron (say the lowest state) in terms of the free representation, we find that it is a superposition of states of a free electron and various numbers of electron-positron pairs. However, in that case we would adopt the point of view that the bound representation is the proper one and gives the appropriate definition of an electron.

The situation is more complicated when we consider more realistic interactions such as that with the electromagnetic field, which is itself quantized (photons can be created and destroyed). Then we distinguish between 'bare' electrons and 'physical' electrons. The bare electrons correspond to the free representation in which interactions are neglected. These states form the starting-point in perturbation treatments of the interaction. The physical electron state is supposed to contain all the effects of the interaction, and in some sense *is* the state which is produced from the bare electron state when the interactions are 'turned on'. It has definite mass and spin, but presumably has an intrinsic structure due to the interactions. This structure consists of a cloud of virtual bare photons and bare electron-positron pairs with one extra bare electron (this statement is more a characterization of the mathematical formalism than a meaningful physical statement). In space this structure is supposed to extend a distance of about $\hbar/mc$, and in time about $\hbar/mc^2$. A physical electron has physical dimensions and other properties such as shape, charge density, and distribution of charge (the last three really mean about the same thing). The 'age' of an electron plays no great role in the theory: i.e. the behaviour of an electron is independent of its age.

This structure of the physical electron due to its interactions has been verified experimentally in many ways: Lamb shift, anomalous moment of electron, effect on hyperfine splitting, and effect on electron scattering from a Coulomb field. The structure comes from a theory in which the fields are local: that is, the intrinsic bare fields are points and the interactions take place at a point. Calculations using this theory give remarkably accurate predictions of the four types of experiments just mentioned—i.e. the internal structure produced by the interactions is well understood. It is conceivable that in addition to this the bare particles could themselves have an intrinsic non-point structure. There is at the moment no evidence for this. However, an important high-energy electron-electron scattering experiment is in progress at Stanford University. With the present theory, the results of that experiment have been predicted in advance. If the experiment agrees with these predictions, it will mean that the electron has no internal structure other than that produced by interactions (I must emphasize that the physical electron has a known structure of dimensions $\sim 10^{-10}$ cm which can be allowed for in the experiment); of course this statement would have been verified only for distances greater than $\sim 10^{-14}$ cm. Disagreement would mean that the laws of physics at distances less than 10^{-13} or 10^{-14} cm may be different from our present conceptions. Future experiments will investigate this possible 'breakdown of quantum electrodynamics' in even smaller regions of space-time.

To summarize:

(i) Intrinsically the non-interacting electron appears to have no structure at distances greater than 10^{-13} cm. The Stanford experiment will check whether this is true at smaller distances.

(ii) However, the interactions with the quantized electromagnetic field which lead to electron scattering, pair creation, etc., also give the physical electron an internal structure. This structure can (in principle) be computed accurately, and extends to distances of order $\hbar/mc$.

(iii) On a larger scale one may find the uncertainties associated with the manner of formation of the electron's wave packet. The present point of view is that these do not represent the internal structure of the electron.

APPENDIX III

When we describe the state of a micro-particle by its wave properties, we use a wave function, which is generally complex. Let the wave function be ψ, its complex conjugate ψ^*. Then $|\psi|^2 = \psi\psi^*$ is the intensity of the wave. The exact dependency of the Born probability to the intensity is chosen as unity. If P is the probability, then $P = \psi\psi^*$. P is never negative.[1] It is large where ψ is large, and is small where ψ is small, and does not depend on insignificant quantities such as the zero level of potential energy. The probability of finding a particle between x and $x + \delta x$, y and $y + \delta y$, and z and $z + \delta z$ is then

$$P(x, y, z)\,\delta V = \psi\psi^*\,\delta x\,\delta y\,\delta z. \tag{1}$$

The exact form of the wave function is determined through Schrödinger's equation. Let H be the quantum-mechanical Hamiltonian, E the allowed energy levels or eigenvalues, and ψ_n those wave functions that satisfy Schrödinger's equation—which can now be written

$$H\psi_n = E_n\psi_n. \tag{2}$$

The ψ_n are the stationary states, and the equation is in its eigenvalue form. Let us restrict ourselves to one dimension. In classical mechanics, H may be written[2]

$$H = \frac{P^2}{2m} + V. \tag{3}$$

Schrödinger's equation may also be written

$$\left(\frac{-\hbar^2}{2m}\frac{\partial^2}{\partial x^2} + V\right)\psi_n = E_n\psi_n. \tag{4}$$

Comparing
$$H = \frac{P^2}{2m} + V = \frac{-\hbar^2}{2m}\frac{\partial^2}{\partial x^2} + V,$$

$$P^2 = -\hbar^2\frac{\partial^2}{\partial x^2}. \tag{5}$$

Corresponding to the classical momentum p, there is a quantum-mechanical operator P. Then

$$p = m\frac{dx}{dt} \to P = \frac{\hbar}{i}\frac{\partial}{\partial x} \quad \left[i^2 = -1,\ \hbar = \frac{h}{(2\pi)}\right].$$

171

In Schrödinger's equation, H is an operator, operating on the stationary state, ψ_n. Only certain energy levels are allowed for a stationary state (the so-called 'energy levels'), as in the hydrogen atom, where 13.6 eV is the highest non-ionized energy level.

Let us reappraise, qualitatively, our moves in section B thus far. Assuming the energy-frequency law, and ascribing wave properties to particles, and particle properties to light waves, we then describe *matter waves*. We introduce the concept of wave intensity; it is equal to the absolute value of the square of the 'state'. The intensity (and hence the state) is identified with the probability of determining the position of the particle. The equation for determining the 'steady state' has been given: corresponding to the classical momentum there is a *quantum*-mechanical momentum operator, operating on the wave function, or state. The classical Hamiltonian also becomes an operator in the transition to quantum mechanics. The energy eigenvalues were identified with the energy levels. Quantum mechanics can be formulated via operator formalism. And, of course, there is another approach, the matrix method, whose 'equivalence' with the Schrödinger approach (as felt to obtain by most contemporary physicists), will now be demonstrated.

A matrix is an array of quantities introduced in linear transformation theory. Let us say that we are given a vector $V = (x, y, z)$ in space, referred to the axes of a Cartesian co-ordinate system, S. If we move to another co-ordinate system, S', the same vector will have the same absolute length—but its components will be different. Denote the vector (as viewed in S') as $V' = (x', y', z')$. The invariance of length requires

$$V . V = V' . V'.$$

We now seek the relation between V in S and V' in S'. If S differs from S' by a translation alone, we can write $x' = x + a$, $y' = y + b$, $z' = z + c$ (where a, b, c are constants).

For a rotation, the relations are more complicated. Let us work in two dimensions for clarity. Then

$$x' = x \cos \phi + y \sin \phi,$$

$$y' = y \cos \phi - x \sin \phi.$$

If
$$a_{11} = a_{22} = \cos\phi, \quad a_{12} = -a_{21} = \sin\phi,$$
$$x' = a_{11}x + a_{12}y,$$
$$y' = a_{21}x + a_{22}y.$$

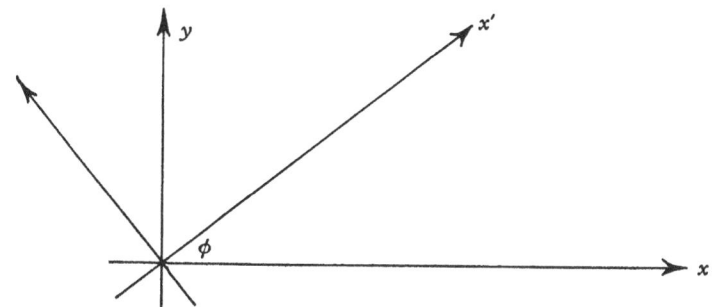

Let us now change the notation. Let (x', y') become (x_1', x_2') and (x, y) become (x_1, x_2). We may then write the above transformations in summation form

$$x_i' = \sum_{j=1}^{h} a_{ij}x_j \quad (i = 1, 2). \tag{6}$$

The quantities a may be arranged as follows:

$$(A) = \begin{pmatrix} a_{11} & a_{12} \\ a_{21} & a_{22} \end{pmatrix},$$

(A) is a 2×2 matrix. The vector may be written in matrix form as

$$(x) = \begin{pmatrix} x_1 \\ x_2 \end{pmatrix} \text{ in } S, \quad (x') = \begin{pmatrix} x_1' \\ x_2' \end{pmatrix} \text{ in } S'.$$

Then (6) becomes
$$(x') = (A)(x).$$

Knowing the result, we can find a rule for matrix manipulation.

$$\begin{pmatrix} a_{11} & a_{12} \\ a_{21} & a_{22} \end{pmatrix} \begin{pmatrix} x_1 \\ x_2 \end{pmatrix} = \begin{pmatrix} x_1' \\ x_2' \end{pmatrix} = \begin{pmatrix} a_{11}x_1 + a_{12}x_2 \\ a_{21}x_1 + a_{22}x_2 \end{pmatrix}.$$

In summation form the matrices, A and B, may be multiplied together to yield a third matrix C. Then,

$$C = AB,$$
$$C_{ij} = \sum_{K} a_{iK}b_{Kj}.$$

Our alternate formulation of quantum theory depends upon the representation of quantum-mechanical operators as matrices. By solving Schrödinger's equation for ψ_n, we get not a single stationary state as a solution, but rather a number of allowed stationary states which form an orthogonal set.

Suppose two members of the set ψ_n are ψ_r and ψ_q. Then the orthonormality condition requires

$$\int \psi_r^* \psi_q dx = \delta_{rq},$$

where δ_{rq} is a symbol following the rules

$$\delta_{rq} = 0, \quad \text{when} \quad r \neq q,$$
$$\delta_{rq} = 1, \quad \text{when} \quad r = q.$$

An example of such a set is $\psi_n = \exp(2\pi i n x/L)$, where the different elements of the set are found by assigning n integral values $(n = 1, 2, \ldots)$. For the following discussion, ψ_n will refer to a complete orthonormal set, ψ_q will refer to the qth element, obtained by letting $n = q$, and similarly ψ_r is the rth element of set ψ_n. A property of our operators is that an operator operating on a wave function yields another wave function $P\psi_r = \phi_r$. We now state an expansion postulate: *wave functions can be expanded as a series of the eigenfunctions of our operators.* This is the consequence of the completeness of the set ψ_n. Then ψ_r should be expandable in a series of $P\psi_r = \phi_r = \sum a_{nr}\psi_n$, where the a_{nr} are the coefficients of the series expansion. Then

$$P\psi_r = \sum_n a_{nr}\psi_n;$$

also
$$\psi_q^* P\psi_r = \psi_q^* \sum_n a_{nr}\psi_n$$
$$= \sum_n \psi_q^* a_{nr}\psi_n$$
$$= \sum_n a_{nr}\psi_q^*\psi_n.$$

Integrating all over space

$$\int_\infty \psi_q^* P\psi_r dx = \int_\infty \left(\sum_n a_{nr}\psi_q^*\psi_n\right) dx,$$

remembering that

$$\int_\infty \psi_q^* \psi_n dx = \delta_{qn} \quad \text{(orthonormality)},$$

$$\int_\infty \psi_q^* P\psi_r dx = \sum_n a_{nr}\delta_{qn} = a_{qr}.$$

The array of numbers a_{qr} (which are generally complex) form a matrix which we shall call the matrix for P, a_{qr} being the q-rth element.

$$\int_{\infty} \psi_q^* P \psi_r dx = (P)_{qr}. \tag{7a}$$

Analogously, for X, the position operator

$$\int_{\infty} \psi_q^* X \psi_r dx = (X)_{qr}. \tag{7b}$$

We are now in a position to solve the simple harmonic oscillator problem by the matrix methods. We will be working in one dimension for the problem. Consider the Hamiltonian $H = (P^2/2m) + V$. For the oscillator, $V = \frac{1}{2}kX^2$, so

$$H = \frac{P^2}{2m} + \frac{1}{2}kX^2.$$

The eigenvalue equation is $H\psi_n = E_n \psi_n$ or

$$\left(\frac{P^2}{2m} + \frac{1}{2}kX^2\right)\psi_n = E_n\psi_n. \tag{8}$$

Now multiply (8) by ψ_r^* and integrate over all space. Now

$$\int_{\infty} \psi_r^*\left(\frac{P^2}{2m} + \frac{1}{2}kX^2\right)\psi_n dx = \int_{\infty} \psi_r^* E_n \psi_n dx$$

$$= E_n \int_{\infty} \psi_r^* \psi_n dx$$

$$= E_n \delta_{rn}.$$

Considering δ_{rn} as the unit diagonal matrix, since the E_n are constants, the eigenvalue matrix for the Hamiltonian, the energy matrix, will be diagonal and will be of the form

$$\begin{pmatrix} E_1 & 0 & 0 & \cdots \\ 0 & E_2 & 0 & \cdots \\ \cdots\cdots\cdots\cdots \end{pmatrix}.$$

We have

$$\left.\begin{aligned} \frac{1}{2m}\int_{\infty} \psi_r^* P^2 \psi_n dx + \frac{1}{2}k\int_{\infty} \psi_r X^2 \psi_n dx = E_n\delta_{rn}, \\ \frac{1}{2m}(P)_{rn}^2 + \frac{1}{2}k(X)_{rn}^2 = E_n\delta_{rn}. \end{aligned}\right\} \tag{9}$$

The equations of motion for the oscillator are

$$P = m\left(\frac{dx}{dt}\right) = m\dot{x}, \tag{10a}$$

$$F = \dot{P} = -kX, \tag{10b}$$

Eliminating P between (10a) and (10b)

$$\frac{1}{m} P = \ddot{X} = \frac{-k}{m} X. \qquad (10c)$$

Defining the vibrational frequency

$$f_0 = \frac{1}{2\pi} \sqrt{\frac{k}{m}}, \qquad (10d)$$

$$\ddot{x} = -4\pi^2 f_0^2 x. \qquad (10e)$$

In matrix form

$$(\ddot{x})_{qr} = -4\pi^2 f_0^2 (x)_{qr}.$$

If ψ_n depends upon the time explicitly, Schrödinger's equation may now be written

$$H\psi_n = i\hbar \left(\frac{\partial \psi_n}{\partial t} \right),$$

$$i\hbar \left(\frac{\partial \psi_n}{\partial t} \right) = E_n \psi_n.$$

Solving for ψ_n

$$\psi_n = \phi_n \exp[-iEt/\hbar] \quad \text{(for } x\text{)}.$$

Then if $(P)_{qr} = \int_\infty \psi_q^* P \psi_r \, dx$, we see that

$$(P)_{qr} = \exp[2\pi i (E_q - E_r) t/h] \int_\infty \phi_q^* P \phi_r \, dx.$$

But

$$\frac{1}{h}(E_q - E_r) = f_{qr}.$$

Denoting the matrix elements $\int_\infty \phi_q^* P \phi_r \, dx$, by b_{qr}, we get

$$(P)_{qr} = b_{qr} \exp[2\pi i f_{qr} t]. \qquad (11a)$$

Similarly

$$(X)_{qr} = a_{qr} \exp[2\pi i f_{qr} t] \qquad (11b)$$

from (10a)

$$(P)_{qr} = m(\dot{X})_{qr} = 2\pi i m f_{qr}(X)_{qr},$$

and

$$(\ddot{X})_{qr} = -4\pi^2 f_{qr}^2 (X)_{qr},$$

but from (10e)

$$(\ddot{X})_{qr} = -4\pi^2 f_0^2 (X)_{qr},$$

so

$$(f_{qr}^2 - f_0^2)(X)_{qr} = 0,$$

since (X) is not to vanish, for physical reasons,

$$f_{qr} = \pm f_0,$$

$$(E_q - E_r) = \pm h f_0.$$

We must evaluate the quantity $(XP-PX)_{qr}$. It is more convenient to do this by operator methods.

$$(XP-PX)\psi = -\hbar i\left[x\left(\frac{\partial\psi}{\partial x}\right) - \frac{\partial}{\partial x}(x\psi)\right]$$

$$= -\hbar i\left[x\left(\frac{\partial\psi}{\partial x}\right) - x\left(\frac{\partial\psi}{\partial x}\right) - \psi\right]$$

$$= \hbar i\psi,$$

$$\int_\infty \psi_q^*(XP-PX)\psi_r\,dx = \hbar i\int_\infty \psi_q^*\psi_r\,dx,$$

$$(XP-PX)_{qr} = \hbar i\delta_{qr}.$$

We now assume that $q = r+1$. Then

$$f_{r+1,r} = \frac{1}{h}(E_{r+1} - E_r) = f_0.$$

Then the matrix for X must be arranged as follows, remembering that the exponential form of the elements of the matrix yield

$$X_{21} = X_{12}^*, \quad X_{32} = X_{23}^*, \quad ..., \quad X_{ji} = X_{ij}^*, \quad ...,$$

so $\quad (X) = \begin{pmatrix} o & X_{12} & o & ... \\ X_{12}^* & o & X_{23} & ... \\ o & X_{23}^* & o & ... \end{pmatrix}$ (assuming the a's are Hermitean).

The matrix for P will have the same form, since

$$(P)_{qr} = 2\pi i m f_{qr}(X)_{qr}:$$

forming $(XP-PX)_{qr}$ with $q = r+1$ by matrix multiplication, equating it to the $i\hbar\delta_{qr}$ the non-vanishing terms yield:

$$f_0\{|(X)_{r,r+1}|^2 - |(X)_{r,r-1}|^2\} = \frac{h}{8\pi^2 m},$$

$$|(X)_{r,r+1}|^2 = (r+1)\frac{h}{8\pi^2 m f_0} \quad (r = o, 1, 2, ...).$$

And for the matrix elements of (x)

$$(X)_{r,r+1} = \sqrt{\left[(r+1)\frac{h}{8\pi^2 f_0 m}\right]}\exp\left[-i(2\pi f_0 t + \xi_n)\right],$$

where ξ_n are arbitrary phase constants. Then

$$(P)_{r,r+1} = 2\pi i m f_{r,r+1}X_{r,r+1},$$

so
$$(P)_{r,r+1} = -i\sqrt{\frac{(r+1)}{2}}\, hf_0 m \exp\left[-i(2\pi f_0 t + \phi_n)\right].$$

Now
$$(H)_{qr} = E_q \delta_{qr},$$

$$(H)_{qr} = \left(\frac{\mathrm{1}P^2}{2m} + \tfrac{1}{2}k(X)^2\right)_{qr},$$

and we find
$$\delta_{qr} E_q = (q + \tfrac{1}{2}) hf_0 \delta_{qr}. \tag{12}$$

This last equation (12) gives the non-vanishing diagonal elements for the energy matrix. These results are the same as those obtained by the Schrödinger method, as they should be.

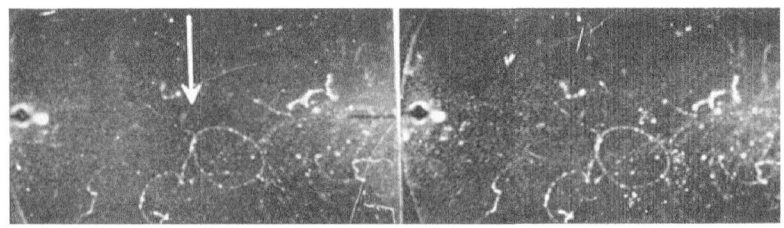

From *Radiations from Radioactive Substances* by E. Rutherford,
J. Chadwick and C. D. Ellis, 1930.

APPENDIX IV

LETTERS FROM PROFESSOR D. SKOBELTZYN

Moscow, 10 October 1956

Dr N. R. Hanson
St John's College
Cambridge, U.K.

Dear Dr Hanson:

Dr Dirac handed me your kind letter and the prospectus of your projected book....

An international colloquium (a conference) was held at Cavendish Laboratory in 1928. I spoke at this conference about my new observations of cosmic ray tracks in Wilson chambers.

The photograph in *Zeitschrift für Physik* (1927, XLIII, p. 362), an enlarged reproduction of which is enclosed in your letter, is certainly a result of near coincidence (overlapping) by chance of a track originating in the gas of the cloud chamber and a spurious track originated without the gamma-rays beam that was under observation. Unfortunately, during the war I lost practically all of my original photographic plates. I have only the print copy of the photo in question in my possession which cannot be examined now stereoscopically with sufficient resolution by means of Pulfrich stereocomparator as it was done during my research in 1927–31. And yet it seems clear to me that if one examines carefully this photo even with poor resolution, using the opaque photo print now on hand (but not its typographical reproduction), one can see that it is a case of near coincidence by chance and not a real pair. But I repeat that this case as well as many others were carefully investigated during my research work by means of Pulfrich stereocomparators and I came to the conclusion that one has to do with a single track originated in the gas of the chamber and not with a double track. Under this last supposition I would interpret the case as single scattering.

In this connection I call your attention to the footnote on page 364 of my article in *Zeitschrift für Physik*, XLIII, 1927.[1]

Dr Chadwick wrote me a letter in which he inquired about details concerning this footnote. In reply to his letter I sent him a photo the reproduction of which you can find on plate VIII (pp. 238–9) of the well-known book *Radiations from Radioactive Substances* by E. Rutherford, J. Chadwick and C. D. Ellis (Cambridge, 1930). On page 239 of this book is given a description and interpretation of the case. This interpretation, which I had in mind at that time, can be judged as a wrong one now. Is it eventually a 'pair' created by gamma-rays of RaC? Even if it is so, I am convinced that it would be completely impossible to establish the fact of pair-creation on the ground of energy-balance only (implying the existence of a new, not known at the time, particle), using in such observations RaC as a source of gamma-rays, owing to the complexity of the spectral composition of these gamma-rays.

The situation is somewhat different in the case of ThC.

In this connection I have to quote my note in *Nature* of January 6, 1934 (v. 133, p. 23).

In this case it is quite clear that as early as in 1931 I observed some cases of positron-electron pairs but gave them a wrong interpretation in the sense just mentioned above as of an energy loss by radiation in a nuclear collision....

During the years 1929–31 I was working at the Institut du Radium in Paris. After my return to the U.S.S.R. I was carrying on a systematic correspondence with Professor F. Joliot-Curie.

In one of his letters (soon after my return) I was told about the lectures that were delivered by Professor R. Millikan in autumn 1931 (or early in 1932) in Paris at the 'L'ecole normale', I think. At these lectures, for the first time (prior to their publication) were reported Dr C. Anderson's observations in the counter-controlled Wilson chamber with a strong magnetic field. Even at that time the positrons were not identified and the corresponding tracks were ascribed (in the above-mentioned lectures) to the protons. In reply to Dr F. Joliot-Curie I immediately informed him that this interpretation cannot be correct. This was quite obvious for me at that time because the ionization density revealed a relativistic velocity of the particles, and protons could not be relativistic in connection with the observed $H\rho$-values. Unfortunately, I have lost Dr Joliot-Curie's letter, but he seems to remember this fact....

I have to limit myself with these brief remarks, but on some other occasion the subject might be discussed more extensively.

With best regards,
Sincerely yours,
D. SKOBELTZYN

Moscow, 22 October 1960

Professor N. R. Hanson
Department of History and Logic of Science
Social Science Building
Indiana University
Bloomington, Indiana

Dear Professor Hanson,

Some time ago I received the draft of a chapter from your work on the history of microphysics.

Now I am sending you back this draft with some notes, as you requested.

I am also sending the enclosed reproduction of the photograph (of 1927) that attracted (in vain!) your attention.

I am trying once more to convince you that the whole story (infra pages 136–139) based on your assertion that the related tracks are tracks of an electron-positron pair is a mistake.

I beg you to examine this photo [facing p. 137] in a stereoscope.

You will see then clearly that only two different interpretations are possible:

Either (1) one has to deal with an overlapping by chance of two different tracks—one of a Compton recoil-electron generated in the gas of the cloud chamber and the other a spurious track emerging from the wall of the cloud chamber (if one examines carefully the stereoscopic picture one sees clearly that this is just the case) or

(2) three particles are produced in the gas of the chamber (one positron and two electrons) so that the positron is emitted in the direction just opposite to the direction of emission of one of the two electrons and that both (positron and its partner-electron ejected in the backward direction) were produced exactly with the same energy.

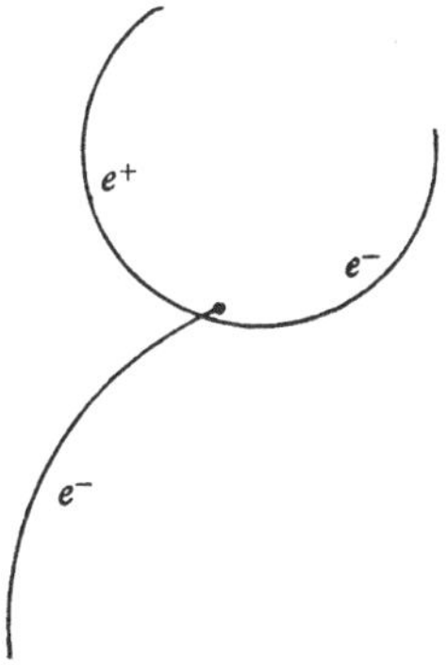

But the latter supposition (the second one) obviously does not make sense and cannot be considered seriously.

Fig. 8

Then only the first one remains and seems to be quite acceptable from statistical point of view.

In a series of observations with ThC (intense line 2620 ekV) which revealed 700 Compton recoil-electron tracks (in a certain angular range) I discovered (in 1931) only 4 electron-positron pairs produced in nitrogen (Note in *Nature*, Jan. 6, v. 133, p. 23, 1934). The statistics of my work of 1927 is based on the observation of only 160 recoil-electrons from RaC γ-rays. The effective positron-producing line (2200 ekV) in RaC γ-rays spectrum is very weak. It is not worth while now to carry out a calculation (which could be done with plenty of theoretical and experimental data now at hand) but I feel the result of such calculation would show that the probability of finding a pair in the course of my observation published in 1927 was small and certainly much less than a unity.

In reference to my letter of 10 October 1956 I must add some remarks.

It seems that it is your intention to publish in your book the excerpts from the above-mentioned letter. I would not mind[1] but on the condition, however, that the following words (from paragraph 3 of my letter) omitted in the draft would be included: 'It is a different question if it were possible for one to establish the existence of positrons only on the ground of energy balance in the observation of that kind. I do not think that it is so.'

Now I shall say that my letter (of four years ago) was written in a hurry (I was anxious to hand it over to Dr Dirac). Later I found in my archives not only the letter of the late F. Joliot-Curie (my reply to him is lost however) but also two other letters related to the same subject— the conferences of R. Millikan in Paris and in Cambridge (autumn 1931).

I am sending you photocopies of the relevant excerpts from these three letters. In addition to F. Joliot-Curie my correspondents were Madame Pierre Curie (Marie Curie) and L. H. Gray—a young physicist from Cambridge (who wrote at that time a long letter about different problems of γ-ray spectroscopy).

I can see from these letters that at that time 'pairs' '$e^+ + e^-$' were already observed by Anderson.

I do not suppose, however, that these documents are of interest to you. But I shall nevertheless point out to you that in case you proposed to use in some way the reference on these letters, the consent of the persons concerned would be needed.

The passage from F. Joliot-Curie's letter is in his own writing and a typewritten copy of it is added for your convenience.

In conclusion I come back again to your point that the positrons could be discovered (and indeed were observed) long time before Dr Anderson's work. I feel (and I must state this bluntly again) that the emphasis on this point in your book leads to a gross distortion in the presentation of the real historical evolution in this field.

The late C. T. R. Wilson certainly saw the tracks of cosmic-ray particles long before my experiments without being able to identify them (and one can find the relevant cases in his beautiful collection of photos published in 1922 or 1923). A simple device (a magnetic field combined with Wilson chamber) used in my experiments for other purposes revealed the nature of these tracks as connected with cosmic rays.

I keep many recollections from this period which may have historical interest, but lies outside the scope of the problem you are outlining in the relevant chapter of your book.

The results of my observations with a relatively weak magnetic field paved the way for the outstanding work of C. Anderson who used a much stronger magnetic field—the main prerequisite of his future success.

It is true that as early as in 1931 (but not earlier as you suggested) I, prior to others, observed electron-positron pairs not being able to identify them, however. (My note in *Nature*, v. 133, p. 23, 1934.) But no one else contrary to what you state (page IX–4) did it at that time or earlier.

At the same time C. Anderson, however, had already obtained his first results. It was his excellent experimental technique that in due course led him to his great discovery.

But in achieving it he was perhaps somewhat slow. Here your psychological considerations are quite plausible.

Sincerely yours,

D. V. SKOBELTZYN

24 October 1960.

NOTES

PAGE 5

1 'Direct action at a distance is out of the question. We cannot conceive of energy disappearing at the sun and reappearing at the earth after an interval of eight minutes without having been propagated continuously in the interval through the intervening space.' Preston, *Theory of Light*, p. 14.

2 For example, Rayleigh's Law of the spectral distribution of radiant energy in thermal equilibrium; this is also contradicted by experimental results at very high frequencies.

PAGE 6

1 Cf. *Verh. dtsch. phys. Ges.* II (1900), 237; *Ann. Phys.* IV (1901), 553.

2 *Ann. Phys.* IV (1905), 17. For Einstein, any ray of frequency v can be considered to be formed of corpuscles containing energy hv, i.e. the higher the frequency of the radiation the greater the *kinetic* energy of the photo-electrons. This runs counter to classical wave theory, where increasing light intensity also increases the amplitude and hence the energy, of the light wave. Einstein's theory puts the energy proportional only to the frequency of the wave, not its intensity (i.e. wave amplitude). This is a fundamental conceptual change.

3 The analogy is inexact because the photons lose their entire energy during collision with matter, and are wholly annihilated.

4 *Phys. Rev.* XXI (1923), 483; *Phil. Mag.* XLVI (1923), 897; cf. also Debye, *Phys. Z.* XXIV (1923), 161.

PAGE 8

1 Cf. Wilde, *Geschichte der Optik* (Berlin, 1838), vol. I, pp. 10ff.

2 *Euclidis Optica* (Heidelberg, 1895): the important perspectival studies based upon this principle include, among many others, those of Damianus (*Damianos Schrift Über Optik* (Berlin, 1897)); Alexandri Aphrodisiensis *in librum de Sensu commentarium* (Berlin, 1902); Archimedes, *Opera Omnia* (Leipzig, 1910–15); Heronis Alexandrini, *Opera* (Die Katoptrik, Nix-Schmidt, 1900); Ptolemy, *Opera astronomica minora, fragmenta* (Leipzig, 1907); L. da Vinci, *Trattato della pittura*; A. Dürer, *Unterweisung der Messung mit Zirkel und Richtescheit* (1525); G. B. Porta, *Magia Naturalis* (1558), describing the 'camera obscura'; Maurolycus, *Photismi de lumine et umbra* (Naples, 1611); Kepler, *Ad Vitellonem Paralipomena, quibus astronomiae pars optica traditus* (Frankfort, 1604); Kircher, *Ars magna lucis et umbrae* (Rome, 1646).

3 Cf. Aristotle, *De Anima*, II, 7. And cf. Plato's *Timaeus* (Jowett), pp. 538, 539, 561.

4 *Liber de speculis*, chs. II and III.

PAGE 9

1 Of course, Aristotle was a formidable dissenter. Cf. *De Anima*, II, 7: 'Light is then the action (ἐνέργεια) of this pellucid (διαφανές) *qua* pellucid. . . . Thus we have shown light to be neither fire [*vs.* Empedocles and Plato], nor body generally, nor even the effluvium or emanation from any body [*vs.* Democritus]. . .' (cf. Young, *op. cit.* I, 473). It may be argued that the intimate connexion which I have asserted to exist between the ancient corpuscular theory of light and the facts of rectilineal propagation and diffraction is falsified by the famous theory of visual rays. For Aristotle, seeing consisted in emanations from our eyes. They reach out, tentacle-fashion, and touch objects whose shapes are 'felt' in the eye. [Cf. *De Caelo* (Oxford, 1922), 290 *a*, 18; and *Meteorologica* (Oxford, 1923), III, iv, 373 *b*, 2. Also Plato, *Meno* (London, 1869), 76 *c–d* and *Topica* (Oxford, 1928), 105 *b*, 6.] Theophrastus argues that 'Vision is due to the gleaming. . .which [in the eye] reflects to the object' (*On the Senses*, p. 26, trans. G. M. Stratton). Hero writes: 'Rays proceeding from our eyes are reflected by mirrors. . .that our sight is directed in straight lines proceeding from the organ of vision may be substantiated as follows. . . .' [*Catoptrics*, 1–5, trans. Schmidt in *Heronis Alexandrini Opera* (Leipzig, 1899–1919).] Galen is of the same opinion. So too is Leonardo: 'The eye sends its image to the object. . .the power of vision extends by means of the visual rays. . . .' (*Notebooks*, C.A. 135 v.b. and 270 v.c.) Similarly, Donne in *The Ecstasy* writes: 'Our eye-beams twisted and. . .pictures in our eyes to get was all *our* propagation.' This is the view that all perception is really a species of touching, e.g. Descartes' *impressions*, and the analogy of the wax. Cf. '[Democritus] explains [vision] by the air between the eye and the object [being] compressed. . .[it] thus becomes imprinted. . . "as if one were to take a mould in wax". . .' (Theophrastus [*op. cit.* pp. 50–3]). But all this, while compatible with the rectilinear principle, and immensely interesting as a contribution to theories of *vision*, is of little value as a theory of *light* itself. We shall overlook this philosophical episode in the history of optics therefore, and persist in associating rectilinearity with corpuscularity, as it was historically the most natural thing to do.

2 Franciscus Maria Grimaldi, *Physico-mathesis de lumine coloribis et iride* (Bononiae, 1665). He also discovered diffraction: 'Lumen propagatur seu diffunditur non solum directe, refracte ac reflexe, sed etiam quodam quarto modo diffracte.'

3 The ultimate explanation of this is given by Young, *Lectures on Natural Philosophy*, vol. I, pp. 367, 378, 389, 464. Cf. Newton, *Principia*, Book II, Proposition L, Scholium, and *Opticks*, query 28.

4 'Lumen videtur esse quid fluidum perquam celerrime et saltem aliquando etiam undulatim fusum per corpora diaphana' (*Physico-mathesis...etc.*).

5 Young, *A Course of Lectures on Natural Philosophy* (1807).

6 Fresnel, *Oeuvres complètes* (Paris, 1866–70), vols. I and II.

7 Compare Grimaldi: 'Lumen aliquando per sui communicationem reddit obscuriorem superficiem corporis aliunde primus illustratum' (*Physico-mathesis...etc.*, p. 187). Of course, Young, always respectful towards Newton's *Opticks*, even gives some credit for discovering the interference principle to his mighty predecessor; he notes how Newton uses a similar idea in explaining the combinations of tides in the Port of Batsha (cf. I. B. Cohen, *Amer. J. Phys.* VIII (1940), 99 ff.).

PAGE 10

1 *Traite de physique*, tom. IV (Paris, 1816).

2 '...the whole space this way and that...will be filled by the dilated waves...the pulses will dilate themselves on all sides...and therefore will fill up the whole space. And we find the same by experience also in sounds...' (Newton, *Philosophiae Naturalis Principia Mathematica*, book II, proposition XLII, theorem XXXIII). Cf. Preston (1890): 'If the medium be homogeneous and isotropic, a (wave) disturbance is propagated with the same velocity in all directions...' (p. 35). Within contemporary phonon theory, this verdict might have to be qualified (cf. J. Ziman, *Electrons and Phonons* (Oxford, 1960)).

PAGE 12

1 For this principle there never has been the slightest empirical justification. Compare: 'Now there are two methods by which we may communicate energy to a body at a distance.... Either matter has come to us from the source, carrying the energy associated with it, as in the case of pellets fired from a gun, or else the energy has been successively propagated through some medium existing between us and the source..., even the possibility of conceiving some new method is perhaps a speculation of a purely visionary character, and is certainly beyond our grasp at present...' (Preston, *The Theory of Light* (London, 1890), pp. 13, 14). Preston's 'justification' is all there has ever been for this 'principle'.

PAGE 13

1 Cf. *Philosophiae Naturalis Principia Mathematica*, scholium of proposition XCVI, theorem L (book I): '...as lately was discovered by

12

Grimaldi...[the rays of light]...in their passage near the angles of bodies...are bent or inflected round those bodies as if they were attracted to them...'.

PAGE 14

1 A spray of pellets alone, whether reflected from, or attracted towards, the edge of a knife, would not generate periodic fringes on the target. So *Principia*, book I, section XIV (first scholium) already requires ether waves just to explain the 'outer' fringes.

2 Against Hooke, who advocated a form of the wave theory, Newton argues: 'For, to me the Fundamental Supposition itself seems impossible; namely that the Waves or Vibrations of any Fluid, can, like the Rays of Light, be propagated in *Streight* lines, without a continual and very extravagant spreading and bending every way into the Quiescent Medium...' (*Phil. Trans.* no. 88, p. 5089). Cf. also *Philosophiae Naturalis Principia Mathematica*, proposition XLII, theorem XXXIII (book II): 'All motion propagated through a fluid diverges from a rectilinear progress into the unmoved spaces' (and cf. note 13, p. 8 *infra*). Cf. also proposition L, problem XII (book II), scholium: '...since light is propagated in right lines, it is certain that it cannot consist in action alone....As to sounds, since they arise from tremulous bodies, they can be nothing else but pulses of the air propagated through it....' Hooke had written of light as 'a quick and short vibratory motion, propagated in every way through a homogeneous, highly elastic medium in streight lines, like rays from the centre of a sphere' (*Micrographia* (1665), p. 15. Cf. also *Lecture on Light* in posthumous works of Hooke (1705), pp. 76 ff.). In this conjecture, Hooke was preceded by the Jesuit Pardis, whose ideas on the subject were incorporated in *L'Optique diversée en trois livres* (Paris, 1682) of C. P. Ango. The first is mentioned in Huygens' *Traité de la lumière*, p. 18; he explicitly states that light is due to waves in the ether, just as sound is due to waves in the air.

PAGE 15

1 The periodicity seen in 'Newton's rings' gave the same trouble: the simple action of corpuscles moving rectilinearly could not explain them. The 'fits' were needed here as well. Cf. Young, *Lectures*, vol. II, p. 541.

PAGE 16

1 To reconcile all this with his observations on thin plates, Newton was obliged further to suppose the length of each fit to vary as the secant of the angle of incidence. Boscovich (*Theoria Philosophiae Naturalis*, 1763) attributed the fits to a polarity of the luminous molecules, which,

by rotating, presented alternately their different sides to the reflecting or refracting surface. Biot's theory was identical in this respect (*Traité de physique*, tom. IV), and is significant inasmuch as it attempts to explain reflexion-refraction solely by reference to particle-properties, and not with what he reckoned were *ad hoc* conjectures drawn from wave theory.

PAGE 17

1 Cf. *Opticks*, book I, part I, proposition VI, theorem V; and cf. *Principia*, book I, section XIV, propositions XCIV to XCVIII, and scholium.

PAGE 18

1 Newton's rejection of Huygens' theory consisted partly in the latter's being atomistic in no respect, and hence not mechanical. The Dutchman's system seemed merely geometrical and—like the merely geometrical theories of Descartes—consistently contradicted physical principles.
Cf. *Opticks*, question 29: 'Are not the Rays of Light very small Bodies emitted from shining Substances? For such Bodies will pass through uniform Mediums in right lines *without bending into the Shadow*' [contrast this with Grimaldi's discovery, pp. 5 and 6 above].

2 That is, since

$$\frac{\sin r}{\sin i} = \frac{v'}{v},$$

if i be greater than r, then v' is greater than v; velocity in a more refracting medium exceeds that in rare media. This runs head on into the wave-theoretic conclusion of Huygens (*Traité de la lumière*) whose geometrical construction for reflexion and refraction demands that light move more swiftly in air than in water or glass (cf. Preston, *The Theory of Light*, p. 19).
Huygens' theory, incidentally, had its greatest difficulty in accounting for the rectilineal propagation of light. Moreover, although he constructed the wave front for reflexion and refraction, his waves lacked what Young later revealed as the defining characteristic of wave motion—periodicity. Huygens had to deny this last in order to account for light waves crossing one another without interference: '...when rays come from different directions, even those exactly opposite, they cross without interference...[so] we do not see luminous objects by means of particles....Light is then transmitted in some other way, a comprehension of which we may get from our knowledge of how sound moves through the air...' (*Traité...*, 4). But, of course, interference is necessary both to explain rectilineal propagation (which Huygens sought to explain by wave envelopes alone) and the several aspects of colour examined by Newton. All

this was left to Young's theory. Thus, so far as the seventeenth century goes, Newton's hypothesis had the best of things. Certainly Newton shatters the wave theory of Huygens in his Query 28, which remarks the inconsistency in supposing, as Huygens does, both that light consists in uniform pressure and motions throughout a uniform medium, and that the anomalous double refraction of Iceland Spar is to be explained by supposing Rays of Light to have different properties on their different sides (cf. Young, I, 462: '...[Huygens] infers that the undulations...must assume a spheroidical instead of a spherical form'). The refractory properties of Iceland Crystal were to Newton '...inexplicable if Light be nothing else than Pression or Motion propagated through Aether'. The concept of *different sides* was a natural adjunct of the light-particle. It led directly to the idea of corpuscular polarity as developed by Boscovich and Biot. This is not to be confused with light polarization, ultimately explained by Fresnel via Hooke's guess that light vibrations are transverse and not longitudinal. Young himself tripped over this point. But while on all these counts the seventeenth-century corpuscular theory outstripped the seventeenth-century wave theory, the former remained profoundly vulnerable on the question of velocity in different media; in this respect Huygens' theory was, and still is, on the side of the angels.

3 *C. R. Acad. Sci., Paris*, XXX (1850), 551; cf. Fizeau, *Ann. Chim. Phys.* V (1849), 29.

PAGE 20

1 On this analysis the Cartesian theory of light is embarrassed. It requires, with Newton, the acceleration of light as it passes into denser media; yet it supposes vision to be excited by a pressure transmitted instantaneously through an infinitely elastic medium filling all space. Still, it contains nothing analogous to the continuous propagation of waves. More strange still is Descartes' statement that heat is a vibratory motion of hot and luminous (i.e. incandescent) bodies, which causes the instantaneous pressure to be transmitted in all directions—as when we watch bright embers. Yet this cause is never said to communicate its vibrations to the medium through which the pressure is propagated.

PAGE 21

1 Cf. I. B. Cohen, 'The First Explanation Of Interference' (*Amer. J. Phys.* VIII (1940), 99ff.).

2 It is true that Newton recognizes the class of optical phenomena, *inflexions*, of which Grimaldi's 'inner fringes' are but a special case. But he does not stress this recognition. Indeed he writes in Query 28: 'The Rays which pass very near to the edges of any Body, are bent a

little by the action of the Body, as we shew'd above; but *this bending is not towards but from* the Shadow, and is perform'd only in the passage of the Ray by the Body, and at a very small distance from it' (*Opticks*, my italics).

3 E.g. *Elements*, book I, proposition IX; the proof of the theorem that from an external point not more than one perpendicular can be drawn to a given line.

PAGE 22

1 *Matière et Lumière* (Albin Michel, 1937), p. 124.

2 One of the objections to Newton's theory consisted in the observation that any particle moving with the speed of light ought to make its momentum felt on impact with an illuminated surface. No contemporary of Newton had detected this; hence it was necessary to suppose the light-corpuscle to be of but negligible mass. Lebedew (*Congrès Intern. Phys.* II, 133, Paris, 1900), Nichols and Hull (*Proc. Amer. Acad. Arts Sci.* XXXVIII (1903), 559, and Poynting (*Phil. Mag.* (1905), p. 169; *J. Phys.* (1910), pp. 675–71; and with Barlow, *Proc. Roy. Soc.* (1910), pp. 534–46) have all established that such a momentum *is* felt. This 'radiation pressure', however, is a direct deduction of Maxwell's Electromagnetic Theory (*A Treatise on Electricity and Magnetism*, vol. II, p. 792) and has nothing to do with the Newtonian hypothesis, save *per accidens*. Is it not an expositional impropriety to credit such a discovery as a confirmation of one theory, when in fact it was predicted by, and sought by the proponents of, the diametrically opposed theory?

3 Cf. Sir G. G. Stokes writing in 1883: '[Newton's emission theory]... shows that we are not to expect to evolve the system of nature out of the depths of our inner consciousness.... It shows that we are not to attach too great importance to great names...' (*Burnett Lectures on Light*); also '...that light consists of undulations' (rests) 'on evidence quite overwhelming'.

4 It is worth noting (*contra* De Broglie) that Newton's theory of fits is a perfect *petitio principii*—a begging of the question. He explains refraction and reflexion, much as one of Molière's heroes explains why opium puts people to sleep. The latter event is credited to the *virtus dormitiva* of opium, i.e. its soporific properties. [Cf. *Le Malade Imaginaire*, Third Ballet Scene.] Similarly, Newton explains refraction and reflexion by supposing that if a light corpuscle arrives at a crystalline surface on the crest of an ether wave, it has a disposition to be reflected; and when it arrives between such crests it has a disposition to be refracted. 'Every ray of light in its passage through any refracting surface is put into a certain transient constitution or state, which in the progress of the ray returns at equal intervals and

disposes the ray at every return to be easily refracted through the next refracting surface, and between the returns to be easily reflected by it...' (*Opticks*, book II, part III, proposition XII). Then follows a description of 'Newton's rings', the phenomenon to be explained, which exactly fits this prior 'explanation'. Compare also the definition preceding proposition XIII: 'The returns of the disposition of any ray to be reflected I will call its Fits of easy Reflexion, and those of its disposition to be transmitted its Fits of easy Transmission, and the space it passes between every return and the next return, the Interval of its Fits.' And what is this but to say that reflexion and refraction occur because in certain circumstances light particles have a disposition to be reflected or to be refracted? Opium causes sleep because it has a disposition to cause sleep. No modern theory of radiation could allow such inelegancy; so the modern theory cannot be the same as Newton's. Incidentally, the ether theory can be similarly scrutinized: there, the wave-like properties of light are explained in terms of the disposition of an unobservable, hypothetical medium, viz. the ether, to undulate in just those ways in which light in fact undulates. Why either of these theories was ever taken to *explain* light remains a mystery.

5 Cf. Preston, *op. cit.* '...an ingenious exponent of the emission theory, by suitably framing his fundamental postulates, might fairly meet all the objections that have been raised against it....(But)...these necessary postulates endow the corpuscles with the periodic characteristics of a wave motion...(which) alone sufficiently explains the phenomena. Hence the one remaining argument against the supposition of corpuscles is that they are superfluous...' (p. 21). Preston wrote three years after the photo-electric was discovered by Hertz, but fifteen years before Einstein explained it. Such 'an ingenious exponent of the emission theory' might have been Arago, whose fertile mind was forever dodging objections to the 'pure' Newtonian theory.

PAGE 23

1 Thus even Preston writes: '...no direct test such as has been supposed to be given by the law of refraction in regard to the velocity of light, or by interference phenomena, can decide between the rival hypotheses' (p. 21).

2 For, as I. B. Cohen makes clear, Newton '...had violated one of the major canons of nineteenth-century physics, which held that whenever there are two conflicting theories, a crucial experiment must always decide uniquely in favor of one or the other' (*Preface* to Newton's *Opticks*, Dover, 1952, p. x). Suppose this canon had not obtained at all.

PAGE 25

1 'The function of general laws in history', *J. Phil.* XXXIX (1942), 35–48, see p. 38.

PAGE 26

1 Cf. my article 'Aristotle's Spheres', forthcoming.

2 '...it is no part of the business of an astronomer to know what is by nature suited to a position of rest, and what sort of bodies are apt to move, but he introduces [geometrical] hypotheses under which some bodies remain fixed, while others move, and considers to which hypotheses the phenomena then actually observed in the heaven will correspond...'. Geminus, quoted in Simplicius, *Commentary on Aristotle's Physics* (ed. by Diels, 1882), pp. 291.21–292.31.

PAGE 27

1 Had Neptune not existed, Newtonian mechanics would have been exposed as inadequate to explain the aberrations of Uranus.

2 Cf. U. J. J. Leverrier, in *Annales de l'Observatoire de Paris*, '*Memoires*', vols. V, XI, XIII, XIV, XV, XXVIII (Gauthier-Villars, 1859–1910). Cf. also F. Tisserand, *Traité de Mécanique Céleste* (Gauthier-Villars, Paris, 1896), vol. IV, ch. XXIX, pp. 521ff. Cf. also my paper 'Leverrier: The Zenith and Nadir of Newtonian Mechanics' (in *Isis*, LIII, (Sept. 1962), pt. 3, no. 173).

PAGE 28

1 'Studien über das Gleichgewicht der lebendigen Kraft zwischen bewegten materiellen Punkten', *S.B. Akad. Wiss. Wien*, LVIII (8 Oct. 1868); *Ann. Phys., Lpz.*, LIII (1894), 959; *Vorlesungen über die Principe der Mechanik*, Leipzig, 1897–1904; cf. also *Populäre Vorträge*, p. 362.

PAGE 29

1 For example, the emission of a β particle from a radioactive substance, or the scattering of a γ-ray photon by an electron.

PAGE 30

1 Cf., for example, D. Bohm, 'A suggested interpretation of the quantum theory in terms of "Hidden" variables', *Phys. Rev.* LXXXV (1952), 166; *Causality and Chance in Modern Physics* (London, 1957); and compare De Broglie, *C.R. Acad. Sci., Paris*, CLXXXIII (1926), 447 and CLXXXIV (1927), 273. Cf. also R. Mould, papers on the concept of 'state' forthcoming in *Annals of Physics*.

PAGE 31

1 Ruling out, of course, the degenerate cases discussed by Heisenberg, *The Physical Principles of the Quantum Theory* (Chicago, 1930), p. 20;

and Margenau, *The Nature of Physical Reality* (New York, 1950), pp. 376–7.

2 Bohm's was no genuine alternative in 1952. In order to get his 'modified' mechanics to work like a quantum theory at all, he takes all those formal features of the current system which offend his deterministic sensibilities and buries them in an obscure technical invention, the 'quantum potentials'. (Cf. also Heisenberg, in *Niels Bohr and the Development of Physics*, London, 1955.)

PAGE 33

1 Mill's Method of Difference is to the point here. Since nothing distinguishes the context in which *P* occurred, from the context in which *P* did not, nothing caused *P*.

PAGE 40

1 By a postdiction I mean simply the logical reversal of a prediction. If a prediction consists in working from initial conditions through boundary conditions to a statement about some future event *x*, then a postdiction will consist in inferring from a statement about some present event *x*, through the boundary conditions, back to already *known* initial conditions. Every prediction, if inferentially respectable, must possess a corresponding postdiction. This is part of Hempel's thesis, and it is sound, necessarily. Adolf Grünbaum has strengthened and clarified Hempel's exposition.

PAGE 42

1 Pages 42–48 are a development of pages 119–126 in my book *Patterns of Discovery*: they function here in a different way.

2 Even this may help. Were one undecided between an atomic and an opposing theory, e.g. that of Descartes, it might help to learn of these unit-constituents of chlorine. Cf. 'Ac proinde si quaeratur quid fiet, si Deus superat omne corpus quod in aliquo vase continetur, et nullum aliud in ablati locum venire permittat? respondendum est, vasis latera sibi invicem hoc opso ofre contigua. Cum enim inter duo corpora nihil interjacet, necesse est ut se mutuo tangant, se manifeste repugnat ut distent, sive ut inter ipsa sit distantia, et tamen ut ista distantia sit nihil; quia omnis distantia est modus extensionis, et ideo sine substantia extensa esse non potent.' (Descartes, *Principia*, II, 18; cf. also *Les Méteores*, Oeuvres VI, 238; Mouy, *Le development de la physique cartesienne*, pp. 101–6.)

3 'The parts of all homogeneal Bodies which fully touch one another, stick together very strongly. And for explaining how this may be, some have invented hooked Atoms, which is begging the Question...' (Sir I. Newton, *Opticks*, vol. III, part I).

NOTES TO PAGES 42-44

4 '...the only view compatible with this picture [of a bricklike structure for crystals] would be that the single bricks themselves possess these properties [of complete crystals], which does not solve the problem but only pushes it one step farther back...' (L. A. Seeber, in Gilbert's *Ann. Phys.*, *Lpz.*, xvi, 16, 229).

5 '...We may indeed suppose the atom elastic, but this is to endow it with the very property for the explanation of which...the atomic constitution was originally assumed....' '...it is in questionable scientific taste, after using atoms so freely to get rid of forces acting at sensible distances, to make the whole function of the atoms an action at insensible distances...it merely transfers the difficulty to the primitive atoms...' (J. Clerk Maxwell, *Atom*, in *Scientific Papers*, vol. II, pp. 471, 480).

6 'This attribution of the properties of crystals to the units of which crystals are composed does not solve the problem but only pushes it one step farther back...' (M. von Laue, Introduction, *International Tables for X-ray Crystallography*, vol. I).

7 '...assume matter to be made up of...small constituent parts and ...postulate laws for the behavior of these parts, from which the laws of the matter in bulk could be deduced. This would not complete the explanation, however, since the question of the structure and stability of the constituent parts is left untouched...' (Dirac, *Quantum Mechanics*, p. 3). And compare: '...should we find that the electron has a complex structure...then such speculation must be pushed one stage farther.... But even if such a state of affairs were to arise, giving a situation very similar to that which arose when it was discovered the atom had a complex structure, we should merely have pushed the frontier one stage farther back...'. (G. K. T. Conn, *The Wave Nature of the Electron*, p. 69).

PAGE 43

1 Let us begin here, despite Strabo: 'And if one must believe Poseidonius, the ancient dogma about atoms originated with Moschus, a Sidonian, born before the Trojan times' (book XVI, II, 24, Loeb Library edition, VII, 271). Cf. Cudworth: '...that *Ancient Atomick Physiology*...was no Invention of Democritus nor Leucippus, but of much greater Antiquity: not only from that Tradition transmitted by Posidonius the Stoick, that it derived its Original from one Moschus a Phoenician, who lived before the Trojan Wars...' (*True Intellectual System of the Universe*, The Preface to the Reader).

PAGE 44

1 Democritus, in Diels, *Fragmente der Vorsokratiker* (A 49, A1, 45). Cf. also Plato: 'Properties such as hard, warm, and whatever their

names may be, are nothing in themselves...' (*Theaetetus*, 156e); Galileo: 'White or red, bitter or sweet, noisy or silent, fragrant or malodorous, are names for certain effects upon the sense organs' (*Opere*, IV, 333 ff., also *Two New Sciences*, pp. 40, 48, 112); Descartes, *Principia* and *Traité de la Lumiere*; Locke: *Enquiry Concerning Human Understanding*, II, ch. 8, 15-22; Leibniz, *Philosophische Schriften*, vol. VII, p. 322; Heisenberg, *Nuclear Physics*, p. 4.

2 Cf. Copernicus: '...those tiny and indivisible bodies called atoms... are not perceivable by themselves...' (*De Revolutionibus*: book I, VI). Cf. also Epicurus, *Letter to Herodotus*: in *Stoic and Epicurean Philosophers* (Modern Library edition), pp. 7-11.

3 Support did not all come from one source. In chemistry, the explanation of why two elements form a compound without residue only when mixed in discrete mass proportions gradually became 'atomic'. (If elements consist of particles of definite mass proportions, and forming a compound involves grouping these, then the laws of definite and multiple proportions are consequences of the theory of the atomic structure of matter.) Faraday's approach was from another direction. He writes: 'Equivalent weights of bodies are simply those quantities of them which contain equal quantities of electricity.... Or, if we adopt the atomic theory...then the atoms of bodies...have equal quantities of electricity naturally associated with them.' For Helmholtz, too (1881), Faraday's laws of electrolysis suggested the existence of atoms of electricity.

4 Ostwald, Mach and Pearson were strict. Ostwald altogether banished the concept of atom from physical science. Mach wrote that: 'The atomic theory is [merely] a Mathematical model used for the representation of facts.' Also, '...atoms, electrons, and quanta are only links (auxiliary concepts) to represent a connected system of science...' (*Science of Mechanics*, pp. 590 ff.). Vaihinger argues: 'Simple atoms [entities without extension]...cannot be actual things' (*Die Philosophie des 'Als Ob'*, p. 219). Pearson puts it that 'Atom and molecule are *intellectual conceptions* by aid of which physicists classify phenomena' (*Grammar of Science*, p. 85). Cassirer: '...a concept like that of material point...can never be understood as the copy of a physical object...' (*Determinismus und Indeterminismus in der modernen Physik*, pp. 164 ff.).

PAGE 45

1 H. Nagaoka, *Nature, Lond.*, LXIX, 392-3, 25 February 1904. E. Rutherford, 'The scattering of α and β particles by matter', *Phil. Mag.* 1911, p. 688. Bohr's theory of the hydrogen atom was a space-time atomic model involving a mechanical orbit for the particle.

2 Cf. '[The Democritean atoms are]...*by hypothesis* the result of division to the last possible stage...hence they cannot in themselves undergo any of the changes experienced by sense-perception...' (K. Freeman, *The Pre-Socratic Philosophers*).

3 Cf. L. De Broglie, *Revolution in Physics*, p. 33; P. A. M. Dirac, *Quantum Mechanics*, pp. 1, 2, 35; W. Heitler, *Elementary Wave Mechanics*, p. 1, lines 1–7. Cf. my article 'The Dematerialization of Matter' in *Philosophy of Science*, 1962.

4 '...the true atoms (in the sense of the ancients) are the elementary corpuscles, for example electrons, which today are considered (perhaps tentatively) as the ultimate constituents of atoms and hence of matter' (L. De Broglie, *Revolution in Physics*, p. 60).

5 Cf. Dirac, *Quantum Mechanics*, pp. 9, 14.

PAGE 46

1 Other methods are the simple ionization chamber (electroscope), the point counter (geiger), the tracer (for neutrons), and more recently the bubble chamber. 'The existence and properties of the ultimate elements are only to be inferred indirectly from observations of gross matter, e.g....as in Millikan's experiments...' (G. Temple, *The General Principles of Quantum Theory*, p. 23). Cf. C. T. R. Wilson's important paper, *Proc. Roy. Soc.* A, LXXXV (1911), 285.

2 After a study of these [particle tracks] one can no longer have the slightest doubt that very small particles have actually flown through space...' (Heisenberg, *op. cit.* p. 33). Cf. R. A. Millikan, *Electrons* (+ *and* −), *Protons, Photons, Neutrons, and Cosmic Rays*; this is the *opus classicus* of the particulate nature of the electron. The wave nature of the electron was established by C. Davisson and L. H. Germer, *Phys. Rev.* XXX (1927), 707; G. P. Thomson, *Proc. Roy. Soc.* A, CXVII (1928), 600; Rupp, *Ann. Phys., Lpz.*, LXXXV (1928), 901; Kikuche, *Jap. J. Phys.*, V (1928), 83; Ponte, *C.R. Acad. Sci., Paris*, CLXXXVIII (1929); the wave nature of particles of atomic mass was established by F. Knauer and O. Stern, *Z. Phys.* LIII (1929), 786; I. Estermann and O. Stern, *op. cit.* LXI (1930), 115. Cf. also T. H. Johnson, *J. Franklin Inst.* CCVI (1928), 301; Ellett, Olson and Zahl, *Phys. Rev.* XXXIV (1929), 493.

3 Merely philosophical uneasiness about, for example, the uncertainty relations, is by itself no good reason to doubt that electrons do have those very properties which entail these relations. Einstein, Bohm, De Broglie may just be philosophically misled. There are shortcomings in present quantum theoretical concepts, without doubt, but these concern algorithmic inconsistencies, and not simply diffuse hunches that God isn't a dice-player.

4 '[Quantum mechanics] requires the states of a dynamical system and the dynamical variables to be inter-connected in quite strange ways that are unintelligible from the classical standpoint...' (Dirac, *op. cit.* p. 15). Remember, even the tentative microphysical explanations within classical physics—those which turn on Newton's ideas of punctiform atomic particles—are far from easy to comprehend. It might even be said that Newton's punctiform masses, in order to meet their assigned tasks within classical physics, must be wholly lacking in classical properties.

5 Here are some examples: '...similar effects [to those which are observed with water waves] take place whenever two portions of light are thus mixed; and this I explain by the general law of the interference of light...' (Young, *Miscellaneous Works of The Late Thomas Young*, I, p. 202); and Rutherford: 'Considering the evidence as a whole, it seems simplest to suppose that the atom contains a central charge distributed through a very small volume' ('Scattering of α and β particles by matter', *loc. cit.* p. 687); and again in 1919: '...it is difficult to avoid the conclusion that the long-range atoms arising from collision of α particles with nitrogen are not nitrogen atoms but probably atoms of hydrogen, or atoms of mass 2...' ('Collision of α particles with light atoms (1919), *loc. cit.* p. 581); '...it was known that many spectral lines consisted of two very close lying components.... In 1925 this doubling was explained [by Goudsmit and Uhlenbeck] in terms of a new property of the electron, *its spin*' (Humphreys and Beringer, *First Principles of Atomic Physics*, p. 279); and Dirac: '...we are led to infer that the negative-energy solutions of [equation 56, *op. cit.* p. 272] refer to the motion of a new kind of particle having the mass of an electron and the opposite charge [the positron]...' (*Quantum Mechanics* (1930), p. 273); and Anderson: '...the tracks shown in Fig. 1 were obtained, which seemed to be interpretable only on the basis of the existence in this case of a particle carrying a positive charge but having the mass of...a free negative electron...' ('The Positive Electron', *Phys. Rev.* (1933), p. 491); and Chadwick: 'The experimental results were very difficult to explain on the hypothesis that the beryllium radiation was a quantum radiation, but followed immediately if it were supposed that the radiation consisted of particles of mass nearly equal to that of a proton and with no net charge, or neutrons'...'The simplest hypothesis one can make about the nature of the particle is to suppose that it consists of a proton and an electron in close combination, giving a net charge 0 and a mass which should be slightly less than the mass of the hydrogen atom' ('Existence of a Neutron', *Proc. Roy. Soc.* A (1932), pp. 694, 700). [Carrying no charge, the neutron leaves no track in a cloud chamber. This made its detection by means of a boron counter difficult, and insured that evidence of it would be 'circumstantial']; Yukawa: 'The interactions of elementary particles are described by considering a hypothetical quantum which has the elementary charge and the proper mass and which obeys Bose's statistics. The interaction of such a quantum with the heavy particle should be far greater than that with a light particle in order to account for the large interaction of the neutron and the proton as well as the small probability of β disintegration. Such quanta [later called "mesons"], if they ever exist and approach the matter close enough to be absorbed, will deliver their charge and energy to the latter....The massive quanta may also have some bearing on the shower produced by cosmic rays...' ('On the interaction of elementary particles', *Proc. Phys-Math. Soc. Japan*, 1935). [Anderson, again, experimentally detected the meson (the heavy, or τ meson) (*Phys. Rev.* LI (1937), 884]; and Fermi: '...the conservation of momentum in the case of the electromagnetic interaction is a necessary condition in order to have a non-vanishing matrix element...' (*op. cit.* p. 20); and Frisch: 'For neither of those assumptions [of the independent-particle model of the atomic nucleus] a good argument could (or can) be made. But once you accept them, a number of dimly noticed regularities fall into place; others were foretold and duly verified...' (*op. cit., loc. cit.*). [Query: but is not this in fact a good argument?] Compare: 'The wave field of an electron...is a probability [amplitude]...without the ψ function no laws of nature could be formulated in the domain of atomic physics' (Heitler, *op. cit.* p. 14). '...Its (ψ) role is mathematical, but for that reason none the less vitally important for the description of nature' (*ibid.* p. 42). '...is linked up with physically observable quantities through a rather long chain of abstractions' (O. Halpern and H. Thirring, *The New Quantum Mechanics*, p. 43). Such 'backward inferences' do not always lead to syntheses like those of Newton, Clerk Maxwell, Einstein and Dirac. They sometimes show only the first small chink in the old armour: when that good Newtonian, Leverrier, 'explained' Mercury's precessions by the non-existent planet 'Vulcan', he did so via an inferential move which had proved monumentally successful in the case of Uranus and Neptune.

PAGE 47

1 For example, to explain surprising observations in series spectra and the anomalous Zeeman effect, Uhlenbeck and Goudsmit (*Naturwissenschaften*, XIII (1925), 953) suggested that the electron was a spinning particle. This proposal was built by Dirac (*Proc. Roy. Soc.* CXVII, 610; CVIII, 351; 1928) into a relativistically invariant theory, yielding spin properties without additional assumptions, and explaining Sommerfeld's fine-structure formula (series spectra) and the separations and intensities in the Zeeman effect (doublet atoms).

Dirac regarded spin as a fundamental property, a premiss of the entire theory, not an accidental adjunct.

2 *Elementary Particles*, p. 2.

PAGE 48

1 Cf. the fundamental paper of Fermi, 'Versuch einer Theorie der β-Strahlen', *Z. Phys.* LXXXVIII (1934), 161. Cf. Heisenberg, *Nuclear Physics*, pp. 122–4. The actual experimental discovery of the Neutrino is reported by Dr Reines and Dr Cowan—from the Savannah River Research Station (*Nature, Lond.*, 1956, *The Neutrino*). Cf. also H. R. Crane, *Rev. Mod. Phys.* XX (1948), 278, which summarizes all neutrino detection attempts up to 1948. L. Langer of Indiana Univ. has also worked extensively in this area.

2 Cf. E. Heisenberg, *Philosophical Problems of Nuclear Science*, pp. 101, 105. Cf. Maxwell writing in 1855: '...the student must make himself familiar with a considerable body of most intricate mathematics.... The first process therefore...must be one of simplification and reduction of the results of previous investigation to a form in which the mind can grasp them. The results of this simplification may take the form of a purely mathematical formula....' 'On Faraday's lines of force, *Trans. Camb. Phil. Soc.* X, 1.

PAGE 49

1 This would be like according to the limit of a series, properties of the series' members; a notoriously unsound procedure. Cf.: 'There is an entirely new idea involved, to which one must get accustomed and in terms of which one must proceed to build up an exact mathematical theory, without having any detailed classical picture.' (Dirac, *op. cit.* p. 12.) '...the main object of physical science is not the provision of pictures, but is the formulation of laws governing phenomena and the application of these laws to the discovery of new phenomena... whether a picture exists or not is a matter of only secondary importance. In the case of atomic phenomena no picture can be expected to exist in the usual sense of the word 'picture' by which is meant a model functioning essentially on classical lines....' (Dirac, *op. cit.* p. 10.) '...to interpret nature on engineering lines proved equally inadequate...to interpret nature in terms of the concepts of pure mathematics [is]...brilliantly successful.' (Jeans, *Mysterious Universe*, p. 143.) '...*all* the pictures which science now draws of nature, and which alone seem capable of according with observational fact, are *mathematical* pictures'...'the universe begins to look more like a great thought than like a great machine....' (Jeans, *op. cit.* pp. 135, 143.) '...the truly creative principle (for physics) resides in mathematics.' (Einstein, Herbert Spencer Lecture, 1933.) '...we have

reached the limits of visualization... the concept of electrons circling a nucleus cannot be taken literally....' (Heisenberg, *Nuclear Physics*, p. 30.) 'This picture of the spinning electron as a rotating ball must not be taken literally. No physical reality whatsoever can be attached to the "structure" of the electron... questions of what the "radius" of such a ball would be, etc., are void of any physical meaning.' (Heitler, *Elementary Wave Mechanics*, p. 70.) 'If we persist in describing phenomena according to the methods of classical physics by means of space and time, then we must give up our ideas of continuity.... If we wish to retain... continuity... we must give up space-time description.... We must not expect to be able easily to picture, by means of models, the fundamental things of nature.' (Flint, *op. cit.* I, 110.) '... schematic idealizations [pictures, classical models]... are capable of representing certain aspects of things, but they have their limits and cannot incorporate into their rigid forms all the richness of reality.' (De Broglie, *Revolution in Physics*, p. 19.) '... it is very difficult to modify our language so that it will be able to describe these atomic processes, for words can only describe things of which we can form mental pictures, and this ability, too, is a result of daily experience. Fortunately, mathematics is not subject to this limitation, and it has been possible to invent a mathematical scheme— the quantum theory—which seems entirely adequate for the treatment of atomic processes....' (Heisenberg, *Physical Principles of the Quantum Theory*, p. 11.)

2 W. Whewell, *Astronomy and General Physics* (London, 1833), pp. 211–12.

3 *Essay* in *Works* (London, 1727), book II, ch. 8, §23.

PAGE 50

1 Heisenberg, *Die Antike*, vol. VIII.

2 Boyle, *Works* (London, 1744), vol. III, p. 15.

3 Lucretius, *On the Nature of Things* (Oxford), book II, p. 98, lines 703 ff.

4 *Op. cit.* book II, pp. 94 ff., lines 842 ff.

5 Bacon, *Works* (London, 1824), Aphorism xxiii, vol. VIII, p. 222.

PAGE 51

1 Birch, *History of the Royal Society* (London, 1756–7), vol. III, pp. 247 ff.

2 Stumpf, *Über den Psychologischen Ursprung der Raumvorstellung* (Leipzig, 1873), p. 22.

PAGE 52

1 Newton, *Principia*, p. xviii.

2 Euler, *Anleitung zur Naturlehre* (in *Opere Posthuma*, II, Leipzig and Berlin, 1911), vol. VI, §50.

3 Helmholtz, *Über die Erhaltung der Kraft*.

PAGE 54

1 But think again of Heitler's caution: 'No physical reality whatever can be attached to the 'structure' of the electron...questions of what the 'radius' of such a ball would be...are void of any physical meaning'. (*Elementary Wave Mechanics*, p. 70.)

PAGE 55

1 Relevant here are the papers by R. Mould, referred to earlier on p. 30, note 1 (p. 192). Compare also the 'state' terminology used in this book on pages 148 ff.

PAGE 60

1 This chapter begins with a development of pp. 149 ff. of *Patterns of Discovery*.

2 For example, consider the sentence 'The sun rises in the east' uttered in two different contexts. In one, 'east' means 'where the sun rises'; in the other 'east' is determined by celestial co-ordinates, and it is thus contingent that the sun rises *there*. Symbolic identity often conceals differences in logical consequences.

3 Weyl, *Philosophy of Mathematics and Natural Science*, pp. 185–6.

4 Theoretical works (e.g. Weyl's *Gruppentheorie und Quantenmechanik* (Leipzig, 1928), appeal to the Principle in a way which is both legitimate and clear. In the formative days, viz. 1913–27, quantum physicists often used classical mechanics as a criterion of the correctness of their calculations, and as a storehouse of suggestions about research and development within the theory. Bohr continually appealed to the Principle for these purposes. One of his associates worked out a complete quantum-theoretical account of the Stark effect on the basis of this principle (Kramers, Dissertation. *K. danske vidensk. Selsk. Skrifter, Naturvidensk.* Afd. 8, Raekke, III (1919), 3, 287. Cf. also Kramers and Heisenberg, *Z. Phys.* XXXI (1925), where the correspondence principle is used for dispersion problems). Schrödinger and Dirac have often looked to classical physics for new ideas (cf. Dirac, *Quantum Mechanics*, 3rd ed. pp. 84, 85).

PAGE 61

1 Nor in Heisenberg's of 1925–6 (matrices), Bohr's of 1926 (using Landé's q-numbers), Born's (statistical), Schrödinger's, Jordan's or von Neumann's of 1926–30 (all operator calculi). Cf.: '...we know nothing of the possibility of commutating the co-ordinate matrix with the momentum matrix...the following relation must hold for

NOTES TO PAGE 61

them: $pq - qp = (h/2\pi i)$ 1' (Halpern and Thirring, p. 29); 'Only those quantities can be sharply defined simultaneously that are commutative (in the sense of the calculus of operators)...the mathematical criterion that a function should contain the one quantity f conjugate to a quantity p is that it should be non-commutative, that is $(fq)-(qf) = n...$' (*ibid.* p. 197).

See especially the important section (11, 10) in von Neumann, where the logic of commutative operators is discussed: '$PQ-QP$ need not have sense everywhere' (p. 234); '...if p, q are two canonically conjugate quantities, and a system is in a state in which the value of p can be given with accuracy...then q can be known with no greater accuracy than $\eta = h/2\pi:\epsilon...$' (p. 238) (this follows 'mathematically' (p. 233) from the specification of P and Q as 'non-commutative operators'). And see note 164, where it is shown that Heisenberg's characterization of the electron in terms of two non-commuting operators is such that the possibility of the particle being a punctiform mass is either meaningless in the notation, or constitutes a recommendation that the notation be rejected. Bold physicists have made this latter recommendation, but never in a persuasive way. See further '...the characteristic condition for the simultaneous measurability of an arbitrary (finite) number of quantities,...is the commutativity of their operators...' (von Neumann, *Mathematical Foundations of Quantum Mechanics*, p. 229).

'These quantum-mechanical operators M_z, M_y, M_x *cannot be commuted with one another....* Hence, there is no meaning in saying that we can...simultaneously make measurements of M_z and M_y... we cannot regard it as a paradox that the maximum value of M_z^2 does not coincide with M_y^2, but rather as a beautiful and striking confirmation of the line of reasoning on which Heisenberg's Uncertainty Relation is founded...' (Halpern and Thirring, p. 205); '...if ξ and η are two observables such that their simultaneous eigenstates form a complete set, then ξ and η commute'...'when the two observables commute, the observations are to be considered as non-interfering or compatible...' (Dirac, *Quantum Mechanics*, L, 52). Cf. also §25. 'One of the dominant features of this scheme is that observables, and dynamical variables in general, appear in it as quantities which do not obey the commutative law of multiplication...' (p. 84); and finally: '...Heisenberg's Principle...shows clearly the limitations in the possibility of simultaneously assigning numerical values, for any particular state, to two non-commuting observables...and provides a plain illustration of how observations in quantum mechanics may be incompatible...' (p. 98). Cf. also 'Commutation and uncertainty relations of the field strengths', §8 of Heitler, *The Quantum Theory of Radiation*.

1 More strongly, not one of the standard notations for elementary particle physics is such that it can be made to express

$$d/dt(mv_x) = X - (\partial V/\partial x),$$
$$d/dt(mv_y) = Y - (\partial V/\partial y),$$
$$d/dt(mv_z) = Z - (\partial V/\partial z)$$

(where force components X, Y and Z have a potential V). These fundamental equations of classical mechanics are not even formulable in the 'language-games' of elementary particle physics; the symbols do not fit together in this way. Is it suggested that they might one day be so fitted together—that this is only a temporary experimental limitation? But then, exactly what substitute-concept is one being invited to entertain? There is no 'new' account of an electron, other than those of orthodox quantum physics, which cannot easily be demolished by any one of the observational data the wave-particle *explicans* was originally developed to deal with. So what articulable alternative is there to the uncertainty relations entailed by this *explicans*? Absolutely none at present. (See the excellent discussion on page 43 of Heitler's *Elementary Wave Mechanics*.)

2 Another philosophical aspect of Weyl's remark concerns whether it is really true that the relation of quantum mechanics to classical dynamics is similar to the relation of the latter to relativity physics. There are reasons for denying this; the two issues are logically distinct. But the discussion cannot be pursued here. Suffice it to say that there are similarities, but also deep conceptual differences. Just as certain phenomena might be described indifferently by classical or by relativity mechanics (e.g. the kinematical acceleration of Mercury as it approaches perihelion), so also there are phenomena which might be described indifferently by classical mechanics or by quantum mechanics (e.g. the hydrogen atom example given in this paragraph). But such operational indistinguishability is no guarantee of logical compatibility, as the rest of the chapter is intended to argue; in fact, the operational overlap has for too long obscured the basic conceptual issue. Von Neumann remarks the matter (*op. cit.* pp. 325, 326); cf. also Heisenberg, *Physical Principles of Quantum Theory*, p. 2.

1 '...p and q are two incompatible variables...homogeneity with respect to one variable, say q, implies an infinite uncertainty in any incompatible variable p' (Temple, p. 43).

2 Compare: 'Classical mechanics must therefore be a limiting case of quantum mechanics' (Dirac, *op. cit.* p. 84).

3 At least not according to any argument known to the author. Quantum theory is in part a language, with its own formation and transformation rules. Why not say that a cluster of symbols which breaks its rules, as does, for example, '$(d^2r/dt^2)m = F$', expresses no intelligible assertion in the language? Thus '...all the well-known, but not understood, "rules" come out one after the other as the result of...absolutely cogent analysis;...once the hypothesis about ψ has been made, no accessory hypothesis is needed or is possible...' (Schrödinger, *Wave Mechanics*, p. 20). And compare the important passage in Heisenberg: 'Any use of the words "position" and "velocity" with an accuracy exceeding that given by equation ($\Delta x.\Delta v \cong \hbar/m$) is just as meaningless as the use of words whose sense is not defined' (*Principles of Quantum Theory*, p. 15).

4 An analogy with the modern history of logic is tempting. Lewis and Langford's system of strict implication purported both to avoid the 'paradoxes' of the Russell-Whitehead system of material implication, and also to include all of the latter system as a subset of its own theorems. But either strict implication includes material implication, along with its paradoxes, or it avoids these paradoxes by failing to include all of the system of material implication. Or, ' $\supset$ ' is given an interpretation by Lewis and Langford which it did not have in the Russell-Whitehead system, making total inclusion of the system, yet selective avoidance of its 'paradoxes', possible. The latter is, of course, the case. In the issue before us an analogous re-interpretation will be sought, to resolve the conceptual tension between quantum mechanics (whose Uncertainty Principle logically distinguishes it from classical mechanics) and the usual interpretation of the Correspondence Principle (which apparently makes any distinction between the two theories a matter of convenience, not a matter of logic). Compare p. 89.

5 '...Up to (1925)...quantum theory...was a conglomeration of essentially different...and partially contradictory fragments (e.g.)... the correspondence principle, belonging half to classical mechanics and electrodynamics (and)...the self-contradictory dual nature of light...' (von Neumann, *op. cit.* p. 4). Not every trace of these origins was obliterated by the Dirac-von Neumann synthesis, as Weyl's earlier quoted remark seems to reveal.

PAGE 64

1 Professor Sir Harold Jeffreys assures me that J. E. Moyal has surmounted this limitation in the statistical formulation (cf. *Proc. Camb. Phil. Soc.* XLIV (1948), 99-124).

Sir Harold also writes: '...the whole of classical mechanics depends on the existence of such (simultaneous probability) distributions. If we knew only the position of a body at an instant, and nothing about its momentum, we could predict nothing at all about its position at any other instant. Quantum mechanics, if it is to be comprehensive, must be in a position to derive the classical equations of motion as approximations valid for systems containing many atoms; and however this is done, some variables corresponding to the co-ordinates and momenta must persist. To deny that they can have a simultaneous probability distribution is to say that quantum mechanics can never explain why classical mechanics gives the right answers for the motion of the planets.' (*Scientific Inference*, 2nd edn., 1957, p. 218.)

2 From a letter written by Sir Harold Jeffreys; printed here with his permission. Von Neumann considers this same possibility: 'The most obvious first step (after encountering the limitations in the theory) would be to assume that this is an incompleteness,...that there must exist a more general formula embracing this as a special case...' (*Mathematical Foundations*..., p. 211).

PAGE 65

1 Von Neumann continues: '...such a generalization...is not possible...in addition to the formal reasons (intrinsic in the structure of the mathematical tools of the theory) weighty physical grounds also suggest this type of a limitation...' (*ibid.* p. 211). '...everything which can be said about the state of a system must be derived from its wave function...' (p. 196). '...a discontinuous operator can never be made continuous by extension' (p. 149). Compare Heitler, *Wave Mechanics*, p. 10.

2 This is an elucidation of a passage which occurred in the last chapter (III, p. 45 ff.).

3 For instance, an electron whirling around a hydrogen-atom nucleus constituted a charge-in-motion which, according to nineteenth-century electrodynamics, should radiate, and should thereby lose energy and instantly collapse into the oppositely charged nucleus. To which Bohr replied simply: 'It doesn't!', a bold, but conceptually unsettling resolution of a deep perplexity.

PAGE 66

1 Cf., again, Heitler, *op. cit.* p. 18, lines 7–16.

2 '...$PQ-QP$ need not have sense everywhere' (von Neumann, p. 234). But it must have sense everywhere within any one physico-mathematical language. Where $PQ-QP$ ceases to have sense is where one physico-mathematical language ends and another begins, despite the fact that the transition may be gradual, and the further fact that the two languages may even be formally analogous at many points (the development of this point for classical mechanics is given in my book *Patterns of Discovery* (Cambridge, 1958), ch. v).

PAGE 67

1 Cf. *Patterns of Discovery*, chs. I and II.

2 Compare von Neumann's fundamental statistical formula on p. 295 of *Mathematical Foundations*.... And *contra* Reichenbach (*Elements of Symbolic Logic*), there is certainly a conceptual divergence between 'the probability that an α is a β is 1' and 'it is logically necessary that every α is a β'. It is logically necessary that all bachelors are male, but it is artificial and misleading to convey this by saying that the probability of this claim's being true is 1. Consider 'All bachelors weigh less than 40 tons': the probability of this being true is 1; the question of probability simply does not arise where semantical unpacking is concerned.

PAGE 68

1 '...not only is the (simultaneous) measurement impossible, but so is any reasonable theoretical definition...' (von Neumann, *op. cit.* p. 326).

2 Heisenberg opts for the same latitude in another context: '*Wave aspect*: Electron creates field; field acts on another electron. *Particle aspect*: Electron emits photon; photon is absorbed by another electron. Both statements describe the same event.' And again: '*Wave aspect*: Neutron creates field; field acts on proton. *Particle aspect*: Neutron emits electron plus neutrino; electron and neutrino are absorbed by proton' (*Nuclear Physics*, pp. 97–8).

PAGE 69

1 Cf. Fermi, *Elementary Particles*, appendix III, pp. 104–5. But Temple conjectures: '...the Correspondence Principle is only a temporary expedient which must sooner or later be replaced by a more profound study of the nature of microphysical systems (*op. cit.* p. 75). This is plausible in a way in which the same remark made about the Uncertainty Principle could not be.

PAGE 71

1 H. Mehlberg, 'The observational problem of quantum theory', read at the May 1958 meeting of the American Philosophical Association.

2 D. Bohm, *Phys. Rev.* LXXXV (1952), 166, 180; and K. R. Popper, *The Logic of Scientific Discovery* (1959).

3 P. Feyerabend, 'The quantum theory of measurement', in *Observation and Interpretation* (Butterworths Scientific Publications Ltd, London, 1957).

4 Cf. *Principia*, book I, section XIV, propositions XCIV–XCVIII; *Opticks* book I, part I; VI.

PAGE 72

1 Cf. *Principia*, proposition XLII, theorem XXXIII.

2 M. Planck, *Ann. Phys.* IV (1901), 553; *Verh. dtsch. phys. Ges.* II (1900), 176.

3 P. E. A. Lenard, *Ann. Phys.* VIII (1902), 149.

4 Albert Einstein, *Ann. Phys.* XVII (1905), 132; R. Millikan [*Phil. Mag.* XXXIV (1917), 1] indirectly confirms this theory.

5 A. H. Compton, *Phys. Rev.* XXI (1923), 483.

6 L. De Broglie, *Thesis* (Paris, 1924); Erwin Schrödinger, *Ann. Phys.* LXXIX (1926), 361.

7 C. J. Davisson and L. H. Germer, *Nature, Lond.*, CXIX (1927), 558; *Phys. Rev.* XXX (1927), 705; G. P. Thomson, *Proc. Roy. Soc.* CXVII (1928), 600.

PAGE 73

1 But compare ch. VIII.

2 This point must be made with some caution. Electrons, and all particles of rest mass > 0 (especially Fermions), satisfy a conservation rule which photons do not. Perhaps one cannot, therefore, make the wave-particle descriptions *entirely* symmetrical alternatives. It may be well to read this point forward into ch. VIII. For the present, however, it will be subdued.

3 Cf. ch. IV, note 2, p. 68 (see p. 206).

PAGE 74

1 Niels Bohr, *Phil. Mag.* XXVI (1913), 1.

2 A. Sommerfeld, *Ann. Phys.* LI (1916), 1.

3 H. Nagaoka, *Nature, Lond.*, LXIX (1904), 392.

4 E. Rutherford, *Phil. Mag.* XXI (1911), 669.

5 L. de Broglie, *Thesis* (1924); *Phil. Mag.* XLVII (1924), 446.

6 E. Schrödinger, *Collected Papers on Wave Mechanics* (Blackie and Son Limited, London, 1928).

7 M. Born, *Z. Phys.* XXXVIII (1926), 11.

8 W. Heisenberg (Born and Jordan), *Z. Phys.* XXXIII (1925), 35.

9 M. Born, *Z. Phys.* XXXVII (1926), 863.

10 P. A. M. Dirac, *Proc. Roy. Soc.* CXII (1926), 661; CXIII (1926), 621; CXIV (1927), 710; CXVII (1928), 610; CXVIII (1928), 351.

PAGE 75

1 Dating from 1854; referred to me by Dirac.

2 O. Heaviside, *Electromagnetic Theory* (London, 1894–1912), appendix K.

3 W. Gordon, *Z. Phys.* XL (1926), 117.

4 P. A. M. Dirac, *Proc. Roy. Soc.* A, CXXVI (1930), 360.

5 J. R. Oppenheimer, *Phys. Rev.* XXXV (1930), 461, 562.

6 P. M. S. Blackett and G. P. S. Occhialini, *Proc. Roy. Soc.* A, CXXXIX (1933), 699.

7 Carl Anderson, *Phys. Rev.* XLI (1932), 405: the intricacies of this conceptual interplay are set out *in extenso* in chapter IX.

8 E. Segré, *Phys. Rev.* C (1955), 947; CI (1956), 909; CII (1956), 1659; *Nuovo Cim.* Series 10, III (1956), 447.

9 See *Observation and Interpretation*, p. 49.

PAGE 76

1 D. Bohm, *Phys. Rev.* LXXXV (1952), 166, 180.

PAGE 77

1 *Brit. J. Phil. Sci.* VII (1957), 356.

PAGE 78

1 W. Heisenberg, compare his essay in *Niels Bohr and the Development of Physics* (Pergamon Press, London, 1955).

PAGE 79

1 L. Boltzmann, *Vorlesungen über Gastheorie* (Johann Ambrosius Barth, Leipzig, 1910).

PAGE 80

1 W. Gibbs, *Collected Works* (Yale University Press, New Haven, 1948).

PAGE 81

1 P. S. LaPlace, *Essai philosophique sur les probabilités* (Paris, 2nd edn., pp. 3–4).

PAGE 84

1 $\hbar/i \cdot \partial\psi/\partial t = \hbar^2/2m\nabla^2\psi$; i.e. a linear, homogeneous, partial differential equation for the De Broglie wave function $\psi(x, t)$.

2 L. De Broglie, *Thesis* (Paris, 1924); Erwin Schrödinger, *Ann. Phys.* LXXIX (1926), 361.

PAGE 85

1 Erwin Schrödinger, *Four Lectures on Wave Mechanics* (Blackie and Son Limited, London, 1928).

2 Albert Einstein, *Sci. News Lett.*, *Wash.*, XVII (1948); H. Jeffreys, *Scientific Inference* (Cambridge University Press, London, 1957), 2nd edn.

PAGE 86

1 Cf. *Principia*, Conclusion: 'I have not been able to discover the cause of those properties of gravity...it is enough that gravity does really exist, and act according to the laws we have explained.'

2 N. Bohr, *Dialectica*, 1 (1947).

3 Cf. ch. VII, and cf. *Patterns of Discovery*, ch. VI, by the present author.

4 As was demonstrated by Sir Harold Jeffreys in 1908.

PAGE 87

1 A. Landé, *Foundations of Quantum Theory* (Yale University Press, New Haven, 1955).

2 G. Leibniz, *De la Sagesse* (1693): in *Leibniz Selections*, ed. Wiener, 'On Wisdom'.

PAGE 88

1 Cf. Albert Einstein, *Library of Living Philosophers* (Evanston, Illinois, 1949), vol. VII; Erwin Schrödinger, in *Brit. J. Phil. Sci.* III (1952), 109, 233; M. von Laue, *History of Modern Physics* (Academic Press, Inc., New York, 1950); *Naturwissenschaften* XXXVIII (1951), 60; L. De Broglie, *Revolution in Physics* and *La physique quantique restera-t-elle indeterministe?* (Gauthier-Villars, Paris, 1953); H. Jeffreys, *Scientific Inference* (Cambridge University Press, 1957).

2 D. Bohm, *Phys. Rev.* LXXXV (1952), 166, 180; *Causality and Chance in Modern Physics* (D. Van Nostrand Company, Inc., Princeton, New Jersey, 1957); Bopp, *Z. Naturforsch.* IIa (4) (1947), 202; VIIa (1952), 82; VIIIa (1953), 6; Fenyes, *Z. Phys.* CXXXII (1952), 81. Bopp light-heartedly dismisses this symmetry condition which Bohr, Heisenberg and Dirac have always treated as indispensable [*Observation and Interpretation* (Butterworth's Scientific Publications Ltd, London, 1957)]. But not only have the promised alternative systems (exclusively preferential either to particulate or undulatory notions) not yet been developed enough to have had any impact on the thought of practising physicists, it remains difficult even to form a detailed concept of what such an alternative would be like. A possible exception to this judgement is some recent research by Jean Pierre Vigier: cf. his articles in *Nuovo Cimento*, 1959–61, and his book *Structure des micro-objets dans l'interprétation causale de la théorie des quanta* (Paris, 1956). But, although inspiring in its promise, Vigier's work has still to generate numbers useful in predictions of the quantitative sort familiar to experimental physicists. Some recent technical work by Bohm is also reported to me as being worth serious attention.

3 Alexandrow, *Dokl. Akad. Nauk*, LXXXIV (1952), 2.

4 Janossy, *Ann. Phys.* (6) XI (1952), 432.

5 The recent parity experiments revealed no exception to Lorentz invariance. They only exposed the impossibility of extending the 'proper' Lorentz invariance (to continuous transformations), so as to include the discontinuous ('improper') mirroring transformation. In all experiments it has been immaterial whether or not the mirroring process embraced 'charge conjugation'. The parity experiments showed that mirroring without charge *conjugation* (particles-anti-particles) does not lead to invariant results for weak (decay) inter-actions. Present indications are that if mirroring is redefined to include the charge *conjugation* (so that the mirror image of an electron is a positron), then Lorentz invariance does hold. I owe this point to Professor Konopinski. Even so, as intimated before, this symmetry for photons, or electrons, may not be complete. For example, the unit charge of an electron seems not to be subsumable to any 'wave' picture.

PAGE 89

1 H. Mehlberg, in the symposium *Philosophical Problems of Quantum Mechanics*, at a recent meeting of the American Philosophical Association.

2 P. Jordan, *Phil. Sci.* (1949), XVI.

3 *Op. cit.* pp. 215–21. We have already examined parts of this argument in the two previous chapters.

PAGE 91

1 S. E. Toulmin, *The Philosophy of Science* (Hutchinson's University Library, London, 1953).

PAGE 93

1 Cf. *Observation and Interpretation* (Butterworth's Scientific Publications, London, 1957). Cf. also, Bohm's *Causality and Change in Modern Physics* (Routledge and Kegan Paul Ltd, London, 1957), and his articles in *Phys. Rev.* LXXXV (1952), 166, 180. Feyerabend's

papers in the *Brit. J. Phil. Sci.* as well as his article in the forthcoming volume III of the *Minnesota Studies in the Philosophy of Science* are equally worthy of note. Writings by other Copenhagen critics, e.g. Dr Mehlberg, will be referred to where appropriate.

PAGE 95

1 For example, the Einstein–Podolsky–Rosen *gedankenexperiment* disturbs Bohr's publicized position—as does also the fact that the spontaneous decay of a carbon$_{14}$ atom is usually thought of as *a* transition from one pure state to another pure state of the atom, even in the absence of a measuring system or detector.

PAGE 97

1 Cf. *Niels Bohr and the Development of Physics* (Pergamon Press, London, 1955). In his AAAS (1961) paper Professor Hilary Putnam was insufficiently sensitive to this fact.

2 I take it that this does not include Dirac's equation, and does not specify in general terms, and on general grounds, what the Maxwellian operators div E-ρ, div H, etc. *operate on.* (They are usually assumed to operate on ψ functions.)

PAGE 98

1 Cf. Dyson: '...divergences at large moments due to insufficiently rapid decrease of the whole integrand at infinity...have always been the main obstacle to the construction of a consistent quantum electrodynamics...'. *Phys. Rev.* LXXV (1949), 1744–5.

PAGE 100

1 Cf. Dyson again: '...the whole theory [QFT] is built upon a Hamiltonian formalism with an interaction-function which is infinite and therefore physically meaningless' (*op. cit.* p. 1754). And yet again: '...*a posteriori* justification of [mathematically] dubious manipulations is an inevitable feature of any theory which aims to extract meaningful results from no completely consistent premises' (*op. cit.* p. 1753).

PAGE 101

1 *The Principles of Quantum Mechanics* (Oxford University Press, 1930).

PAGE 102

1 Feyerabend assures me that current work by Bohm and himself is not open to this objection. But Bohm's earlier (1952) papers certainly were.

PAGE 103

1 For example, the theories of circular celestial motion, of the direct proportionality between instantaneous velocity and distance fallen, of the unqualified undulatory nature of light, of 'specific heat substance', of continuous emission and absorption of all energy, etc.

PAGE 107

1 What we call a 'superstate' of a microparticle is simply a classical, simultaneous determination of its five punctiform parameters, viz. the positional co-ordinates, x, y, z; the time co-ordinate t; and the energy e—all determined with complete precision.

2 Cf. for example, Heisenberg, *Nuclear Physics*, p. 26, photograph (*a*).

PAGE 109

1 It is sometimes said that the wave packet has been 'reduced'; cf. Hanson, *Amer. J. Phys.* (January, 1959). But this characterization is not always very helpful.

2 Distinguished advocates of this line of argument are Popper, Jeffreys and Braithwaite.

PAGE 110

1 I put the matter this strongly despite the elegant attempt of Dirac himself '...to show how complete the analogy is between the quantum and classical treatments of an ensemble', *Proc. Camb. Phil. Soc.* xxv (1929), 62. The issue concerning the state-descriptions of *elements* within the ensembles remains untouched by his treatment.

2 Reichenbach, 'Ziele und Wege der Physikalischen Erkenntnis', *Handbuch d. Physik*, v, IV (1929), 78; Zilsel, *Erkenntnis*, v (1935), 59; Hanson, *Patterns of Discovery* (Cambridge), 1958, ch. VIff.; and *Amer. J. Phys.* v 27 (January, 1959), pp. 1–15.

3 Feyerabend, in *Current Issues In Philosophy of Science*, p. 384; Condon and Mack, *Phys. Rev.* xxxv (1930), 579.

PAGE 111

1 'Uncertainty', *Phil. Rev.* (1954); 'Elementary particle theory', *Phil. Sci.* (1956); *Patterns of Discovery* (Cambridge, 1958); 'The Copenhagen interpretation of quantum theory', *Amer. J. Phys.* (1959); 'Five cautions for the Copenhagen interpretation's Critics', *Phil. Sci.* (1959); 'The Dematerialization of Matter', *Phil. Sci.* (1962); 'Wave Mechanics and Matrix Mechanics', *Czech. Jour. Theoret. Physics* (1961).

2 Cf. *Phil. Sci.* (October, 1959), pp. 326ff.

PAGE 112

1 Bohr's 'philosophy' (at one time or another) encapsulates all of the following sentiments: Baseball umpire A: 'I calls 'em as I sees 'em'. Baseball umpire B: 'I calls 'em as they are'. Baseball umpire C: 'Until I calls 'em, they just aren't'.

PAGE 113

1 C. Eckart, *Phys. Rev.* XXVIII (1926), 711 ff.; E. Schrödinger, *Am. Phys.* IV, LXXIX (1926), 734 ff.

2 This move may have theoretical consequences (cf. p. 218).

PAGE 116

1 I am grateful to Professor Hilary Putnam, of M.I.T., for commending to me the necessity of the following interjection at this point: When we speak of consequences of a physical theory, it is well to keep in mind that these may be either consequences of the postulates of the theory by themselves, or of these together with so-called 'auxiliary hypotheses' accepted in physics at the time.

PAGE 118

1 The reason? the 'auxiliary hypotheses' already mentioned form an ill-defined and constantly changing set. In any of the senses of 'observational consequence' discussed above (*a*, *b*, or *c*), the sets of such consequences are not mathematically well defined. The apparent exception is the case of a systematic translation scheme which converts each physical theory T_1 into some other, T_2. But, again, this establishes *identity*, not equivalence.

2 'Algorithm' has a strict and a broad sense. 'Algorithm', as a pure, uninterpreted formalism (the Hilbertian ideal), is too strict for our purposes. Euclid's geometry is also an algorithm. It differs from Geometrical Optics in that Euclid treats 'straight line' in a technical, physically uncommitted way. Geometrical optics interprets this term, viz. as 'a pencil of light'. So, Euclidean geometry provides the algorithm which, when so interpreted, constitutes a physical theory. But Euclid's geometry is not a *purely* uninterpreted calculus. It already contains descriptive predicates of an intra-geometrical type; many explicit definitions introduce into *The Elements* a semantical richness in excess of what very strict formalists require. So Euclid's geometry is an algorithm in the broad sense. It is always this broad sense with which we will be concerned here.

PAGE 119

1 Of course, concerning what it was physically possible to observe the two theories always differed. Copernicus, following the ancient

Aristarchus, argued that only the immense distance of the fixed stars prevented the detection of stellar parallax due to the earth's revolution. In principle, all heliocentrists urged that given sufficiently sensitive equipment (like Bessel's 'heliometer' of 1838) this aberration could be observed. Any Ptolemaist would have denied this. Incidentally, for Copernicus the Geocentrist's *punctum aequans* was a conceptual error. Making the centre of stellar distances different from the centre of stellar motions struck Copernicus as hopelessly illogical.

2 These investigators claimed that they required *no* interpretation in Schrödinger's sense at all. This is just a mistake. An interpretation had to be implied if the algorithm was ever to be co-ordinated with observations. The idea of a physical theory being a computor-in-a-black-box dies hard, as we saw earlier. It was a very much alive idea in early 1926 when Matrix Mechanicians denied any interest in the internal meaning of a theory, but were concerned only with the predictions which tumbled out of it.

PAGE 120

1 Thus compare Reichenbach: 'The third phase followed immediately: it consisted in the physical interpretation of the results obtained. Schrödinger showed the identity of wave mechanics and matrix mechanics.' *Philosophic Foundations of Quantum Mechanics* (1948), p. vi. And Eckart, *op. cit.* p. 726: '...the wave mechanics and the matrix mechanics are mathematically identical....'

PAGE 124

1 J. von Neumann, *Mathematical Foundations of Quantum Mechanics* (Princeton University Press, Princeton, 1955). English translation.

2 E. L. Hill, *Phys. Rev.* CIV (1956), 1173; CVII (1957), 877.

3 J. M. Cook, *J. Math. Phys.* XXXVI (1957), 82. J. M. Jauch, *Helv. phys. acta*, XXXI (1958), 127. S. T. Kuroda, *Nuovo Cim.* Series 10, XII (1959), 431.

4 M. N. Hack, *Nuovo Cim.* Series 10, IX (1958), 731. J. M. Jauch and I. I. Zinnes, *Nuovo Cim.* Series 10, XI (1959), 553.

5 N. N. Bogoliubov and D. V. Shirkov, *Introduction to the Theory of Quantized Fields* (Interscience Publishers, New York, 1959).

PAGE 125

1 E. C. Titchmarsh, *Eigenfunction Expansions* (Oxford University Press, New York), vol. I (1946), vol. II (1958).

2 The notion of an 'eigenfunction of position' is a perfectly acceptable one, notwithstanding the qualms some physicists have over Dirac's

delta-function. The 'delta-function' is not really a function at all: it is a well defined distribution—in the sense of Schwartz and Lighthill.

PAGE 126

1 I owe this point to Professor Hilary Putnam, of M.I.T.

PAGE 127

1 Incidentally, were (1) and (2) completely satisfied, this would not establish the equivalence of the two physical theories, but (as noted before) their absolute identity.

2 Cf. Eckart, *op. cit.* p. 726: 'The author agrees, however, that the wave-mechanics is more fundamental than the matrix mechanics, and holds out more hope for an eventual physical interpretation of the results obtained.'

PAGE 128

1 This already constitutes a physical interpretation for the algorithm of Matrix Mechanics. It is not the case that Matrix Mechanics was free of interpretative addenda. Rather, these addenda were so obviously correlated with actual spectrographic and scattering observations that their very familiarity made them seem somewhat less like 'interpretation' than they were. Schrödinger's interpretation was not enough like this. Hence Heisenberg's vaunted 'Positivism' is much overstated. He ought never to have claimed that Matrix Mechanics was interpretation-free, but rather that the interpretation it *had* (and must have had, to be a physical theory at all) achieved what Schrödinger's could not achieve in describing observations.

PAGE 129

1 Or as determining the density of particles within a parallelepiped of an electron beam, or as estimating the likelihood of finding a scintillation on one area of a target plate rather than another.

PAGE 136

1 R. A. Millikan, *Electrons* (Cambridge, 1935), p. 339. Cf. Neddermeyer, Anderson's collaborator in 1932 and 1937, writing in *Phys. Rev.* 15 March 1961: 'However, it is not to discredit theory or theoreticians to point out that the muon, like the positron, was a purely experimental discovery in the sense that it was made entirely independently of any theoretical considerations of what particles should and should not exist...' (p. 1814).

2 These sentiments were disclosed to the author during conversations and correspondence with the physicists mentioned.

PAGE 138

1 '...il nous a paru que le parcours des rayons H projetés par les neutrons émis vers l'arrière, par rapport à la direction des particules α, est plus grand que celui qu'on peut prévoir par le calcul....Un rayonnement très pénétrant s'absorbant par projection de noyaux et émis dissymétriquement par rapport à la direction des particules α incidents....' (I. Curie and F. Joliot, *C.R. Acad. Sci., Paris,* cxciv (11 April 1932), 1232.)

PAGE 139

1 *The Principles of Quantum Mechanics,* Oxford, 1930; 'The quantum theory of the electron', *Proc. Roy. Soc.* A, cxvii (1928), 610; cxviii (1928), 351.

2 Dr Oppenheimer writes: '...I remember talking with Anderson about the possible relevance of his tracks and the Dirac theory. His attitude was that he did not understand the theory, and could not take a responsible view of it: and that therefore his discovery had to be soundly based on experimental material....' [Letter to the author, 12 August 1960.]

3 C. D. Anderson, 'The positive electron', *Phys. Rev.* (1933), xliii, 492.

PAGE 140

1 Cf. *Electrons,* ch. xiv, figs. 54 and 55.

PAGE 141

1 'Cosmic rays disrupt atomic hearts', in *Science Service.*

2 Williams and Terroux, *Proc. Roy. Soc.* A, cxxvi (1929–30), 300.

PAGE 142

1 Cf. Skobeltzyn, *Z. Phys.* liv (1929), 686; this is the first known reference to such branch tracks.

2 Concerning this work of Millikan and Anderson, Professor Skobeltzyn has kindly sent me correspondence of the period from L. H. Gray, M. Curie and F. Joliot-Curie. The relevant excerpts are set out here with the permission of the authors or their heirs or executors:

Dear M. Skobelzyn,   Cambridge, Nov. 27th 1931

On Monday we had a visit from Millikan...he showed eleven photographs of 'good' tracks...which from the sense of their curvature must have been positive particles....In some cases a proton (?) and electron appeared on the same photograph, and with the help of a vivid imagination one *might* conclude that they had a common

FACULTÉ DES SCIENCES DE PARIS

INSTITUT DU RADIUM

LABORATOIRE CURIE
1, Rue Pierre-Curie, Paris (5ᵉ)

Tél. ODÉON 14-80

Paris, le 1ᵉʳ Janvier 1932

Mon cher Skolodgyna

À propos de rayons cosmiques nous avons assisté à une communication du Pr Millikan. Il nous a fait part d'un résultat tout à fait intéressant. Par votre méthode (qui d'ailleurs il n'a pas cité mais que nous lui avons rappelé après sa conférence) ses collaborateurs ont photographiés des trajets de rayon cosmiques dans un champ magnétique élevé (13000 g) et ils ont vu que certaines trajectoires sont dues à des électrons et d'autres à des protons. Il a même projeté un cliché d'ou l'on voit issu du même point un électron qu'un proton d'énergie considérable 50 10⁶ environ. Vous voyez c'est du nouveau qui vous intéresse certainement.

. . .à propos des rayons cosmiques nous avons assistés à une communication du Pr Millikan. Il nous a fait part d'un résultat tout à fait interessant. Par votre méthode (que d'ailleurs il n'a pas citée, mais que nous lui avons rappelée après la conférence) ses collaborateurs ont photographiés des trajets de rayons cosmiques dans un champs magnétique élevé (13 000 G) et ils ont vu que certaines trajectoires sont dues à des électrons et d'autres à des protons. Il a même projeté un cliché d'ou l'on voit issu du même point un electron et un proton d'énergie considérable 50×10^6 eV environs. Vous voyez c'est du nouveau qui vous interessera certainement. . . .

origin...in every case the proton tracks were more dense than those of the electrons. [This was not in fact true. But that Gray should have expressed himself thus is significant for our thesis.—N.R.H.] Their curvature was clearly visible in some cases, though not of course large...the general feeling was that the curvatures could hardly have been spurious....A large proportion of double tracks, as you had previously found, is in itself very significant....

Yours sincerely
L. Harold Gray

Cher Monsieur,   Paris, le 21 Décembre 1931

...Le point le plus important est que sur ces clichés [of Professor Millikan] on voit, non seulement des électrons, mais aussi des protons (déviation en sens inverse), et qu'on a vu, en particulier, des trajectoires d'électron et de proton paraissant partir d'un même point. Les potentiels indiqués par les courbures seraient de l'ordre de 100 millions de volts....

Veuillez agréer, Cher Monsieur, l'assurance de mes meilleurs sentiments.

M. Curie

Mon cher Skobeltzyne,   Paris, le 1er Janvier 1932

...ses collaborateurs [i.e. Millikan's] ont photographiés des trajets de rayons cosmiques dans un champ magnétique élevé (13 000 G) et ils ont vu que certaines trajectoires sont dues à des électrons et d'autres à des protons. Il a même projeté un cliché d'où l'on voit issu du même point un électron et un proton d'énergie considérable 50×10^6 eV environs....

F. Joliot-Curie

And Professor Skobeltzyn himself remarks that as early as 1931 he 'prior to others, observed electron-positron pairs, not being able to identify them, however. (My note in *Nature, Lond.*, v. 133, p. 23, 1934.)' [Letter to the author, 22 October 1960.]

3 In *Science*, LXXVI (September 1932), 238.

4 C. D. Anderson, *Phys. Rev.* XLIII (1933), 491; cf. also 1034.

5 These tracks are geometrically associated as well, and hence almost certainly time-associated. Given the relative infrequency of cosmic-ray tracks in 1931 (1 in 30 exposures), the likelihood of independent events originating from a common centre is vanishingly small.

Compare Anderson, writing in 1933: 'The tracks...were coincident in time as shown by the fact that the diffusion of the ions broadened all the tracks to the same extent....Seven of the tracks

217

NOTES TO PAGES 142–144

were seen to originate at a common point in the upper portion of the chamber.... The sense of curvature... is such as to indicate electrons. ...Protons...of the degree of curvature observed would have energies too low to be consistent with the observed minimum ranges and specific ionization.... Obviously electrons cannot be the immediate ionizing agents...since more than 1000 electrons would have to be assumed to traverse the ion-chamber simultaneously....'
(*Phys. Rev.* XLIII (1933), 369.)

PAGE 143

1 Cf. *ibid.* XLIV (1933), 406.

2 This paragraph is virtually a paraphrase of Dirac's prefatory remarks in the 1928 paper alluded to in the next footnote. When Professor Heisenberg read this paragraph, however, he reacted as follows: '...I cannot quite agree with the first four lines of [§ B].... Here your report neglects completely the fact that at this time the theory of the electronic spin had been well established. The multiplet structure of the atomic spectra did not at that time offer any essential difficulty for quantum mechanics. Both the multiplet structure and the anomalous Zeeman effect were well understood, and Pauli had introduced the Pauli spin matrices with the well known properties: $\mathscr{C}_x \mathscr{C}_y = -\mathscr{C}_y \mathscr{C}_x = i\mathscr{C}_z$, etc. I cannot doubt that Dirac had been led to his discovery by Pauli's paper and especially by the relation $(p_x\mathscr{C}_x - p_y\mathscr{C}_y - p_z\mathscr{C}_z)^2 = p^2$. The essential progress in Dirac's paper was the connexion of Pauli's spin matrices with the Lorentz group. Dirac succeeded in representing the complete Lorentz group by introducing besides the Pauli spin matrices another similar matrix ζ which in some way represents the space reflexions in the Lorentz group. It is the introduction of this ζ matrix which finally led to the dualism between matter and antimatter...' [Letter to the author, 19 September 1960.]

PAGE 144

1 P. A. M. Dirac, 'The quantum theory of the electron', *Proc. Roy. Soc.* A, CXVII (1928), 610. Cf. also Pauli, *Z. Phys.* XLIII (1927), 601; and Darwin, *Proc. Roy. Soc.* A, CXVI (1927), 227.

2 Although written here precisely as it appears on page 611 of Dirac's 1928 paper, scrutiny reveals that this equation (the Klein–Gordon) is improperly formed. Dirac should have put a minus sign after the first square bracket, giving an equation of the form:

$$F = (-W/c + e/cA_0)^2 + (p + e/cA)^2 + m^2c^2 = 0.$$

As they stand, equations (1) and (3) in Dirac's paper are formally inconsistent, a fact apparently unnoticed during these 33 years.

Professor Dirac has confirmed the author's suggested emendation of (1), and of the equation in the middle of page 612. The latter appears unaltered in the quotation below, which begins 'The second difficulty...'. Again, a minus sign after the square bracket is necessary. Compare Notes by Hutten, and Hanson, in *British Journal of Philosophy of Science*, 1962.

3 Gordon and Klein, *Z. Phys.* XLI.

PAGE 145

1 Author's note: that is, these solutions have a 'negative energy'.

PAGE 146

1 Cf. Gordon, *Z. Phys.* LX (1926), 117 ff.

2 The Dirac equation for an electron moving in an arbitrary electromagnetic field of potentials A_0, A_1, A_2, A_3, is thus of the form:

$$F\psi = [p_0 + (e/c)A_0 + \alpha_1(p_1 + (e/c)A_1) + \alpha_2(p_2 + (e/c)A_2)$$
$$+ \alpha_3(p_3 + (e/c)A_3) + \alpha_4 mc]\psi = 0. \qquad (2)$$

3 This in no way minimizes the great achievement of Goudsmit and Uhlenbeck (1925). Their brilliant hypothesis, however, was intended only to 'save the phenomenon', and not to strengthen electron theory. Cf. *Nature, Lond.*, CXVII (1926), 264. Cf. Dirac: 'The αs [in (2)]... may be regarded as describing some internal motion of the electron, which for most purposes may be taken to be the spin of the electron postulated in previous theories...' (*Proc. Roy. Soc.* A, CXVIII, 351). '...We can interpret this result by saying that the electron has a spin angular momentum of $\frac{1}{2}h\rho$, which, added to the orbital angular momentum m, gives the total angular momentum M, which is a constant of the motion...' (*Proc. Roy. Soc.* A, CXVII, 620). And please compare Heisenberg's remarks on spin, introduced in the first note in this section B. Cf. also 'Exclusion principle and spin', by B. L. Van Der Waerden (in *Theoretical Physics in the Twentieth Century*, Interscience, N.Y.), especially pp. 232–3.

4 This is Professor Hans Bethe's remark.

5 At first Dirac simply excluded these solutions, *ad hoc*: 'Since half the solutions must be rejected as referring to the charge $+e$ on the electron, the correct number will be left to account for duplexity phenomena...' (*op. cit.* p. 618). And even in 1930 Oppenheimer writes: '...since the Dirac jumps do not seem to be *directly* responsible for the difficulties with which we are, in this work, most concerned, we shall for the present neglect them' (*Phys. Rev.* XXXV (1930), 461).

PAGE 147

1 In this connexion Dirac has written (to me): 'The world as we see it is not symmetrical between positive and negative electricity, so a theory that makes it symmetrical would seem to be at fault. That is why I tried to cook the negative energy solutions. The lack of symmetry is still not adequately explained.' (Letter to the author, 1 November 1960.)

2 Weyl, *Gruppentheorie und Quantenmechanik*, p. 234 (2nd edition).

3 Schrödinger, *Dtsch. Phys. Berl. Ber.* (1931), S. 63.

PAGE 148

1 Oppenheimer, 'On the theory of electrons and protons', *Phys. Rev.* XXXV (1930), 562.

2 The main reasons why this won't work are that: (1) If it did one would have to give an infinite density to positive electricity; (2) the scattering of soft light by a proton is unaccountable for on the Dirac theory; '...[the Dirac] theory gives equal scattering coefficients for electron and proton...such interaction would affect electron scattering and proton scattering in exactly the same way; whereas the Thompson formula requires the latter to be smaller by a factor proportional to the square of the ratio of the masses'; and (3) there is a numerical discrepancy in the original Dirac supposition; a mean life for ordinary matter of the order of 10^{-10} sec. 'We should hardly expect any state of negative energy to remain empty' (*ibid.* p. 563).

Cf. also Oppenheimer, *Phys. Rev.* XXXV (1930), 939: '...we compute the rate at which electrons and protons should, on Dirac's theory of electrons and protons, annihilate each other'; ...'The mean lifetime of an electron in a proton density N_p protons per unit volume is thus

$$T = (m^2/c^3/16\pi 5e^4 n_p) \sim 5 \times 10^{10}/n_p \text{ sec....}$$

an absurdly short mean lifetime for matter' (p. 943). Cf. also Dirac, *Proc. Roy. Soc.* A, CXXXIII (1931), 61.

PAGE 149

1 Oppenheimer, 'Interaction of field and matter', *Phys. Rev.* XXXV (1930), 461.

2 Dirac, *Proc. Roy. Soc.* A, January 1930.

3 The idea behind the 'hole' theory is first explicitly put forward by Dirac, *Proc. Roy. Soc.* A, CXXVI (1930), 360; and *Proc. Camb. Phil. Soc.* XXVI (1930), 361. In these papers the 'proton-cooking' occurred. For those to whom the conception of a positron as a 'hole in the negative-charge continuum' is difficult, the following *may* help. Imagine a sardine swimming through an infinite sea. This is in itself no feat of thought. But now 'reverse the Gestalt'. Instead of an

infinite sea of water, imagine an infinite sea *of sardines*! They are so densely packed that they form a virtual continuum. These sardines now correspond to the most stable states for electrons; each sardine is thus 'a negative energy particle of very high velocity'. Imagine now one sardine knocked out of its place in the sea. One would perceive a sardine-shaped vacancy in the sardine-sea, a hole which would stand out from the dense background as being something exceptional, just as a small hole in a large blackboard is immediately apparent. Suppose a sardine immediately adjacent to the hole moves into it, and that the next sardine moves into the new hole, and so also with the next, and the next. One *could* describe this as I have done. Or one *could* think of a sardine-shaped hole itself moving through the continuum. The motions of this hole would then be observed as clearly as would have been the motions of the original sardine in its infinite sea of water. And should any other sardine come skimming along the surface of the sea, it will, when it approaches the hole, immediately fall into it. This will remove both that sardine, and the hole it fell into, from further observation. Only the featureless continuum will again be present; the sardine and the hole will have annihilated each other by the former's collapsing into the latter. But another jolt of energy may dislodge another sardine, creating thereby a *pair* of entities—a sardine-hole in the continuum and an 'ordinary' sardine above the sea.

PAGE 150

1 Halpern and Thirring, *The Elements of the New Quantum Mechanics* (London, 1930–31), p. 101. And compare Condon and Mack, *Phys. Rev.* xxxv (1930), 382: '...According to [Dirac's theory] electrons can exist in states of negative energy (less than $-mc^2$), as well as in usual states of positive energy (of the order of $+mc^2$). We have no experimental evidence of electrons of this sort, and so the prediction of such negative energy electrons was regarded as a blemish in the theory....'

2 Oppenheimer, *Phys. Rev.* xxxv (1930), 939.

3 Pauli, *Handb. Phys.* (1933), ch. 2, B. 5: 'Übergänge zu Zuständen negativer Energie, Begrenzung der Diracschen Theorie', vol. 24, pt. 1 (pp. 242 ff.). Professor Skobeltzyn remarks: 'This view [of Pauli's] prevailed in the opinion of our theoreticians up to the discovery of C. Anderson and Blackett and Occhialini' [letter of 22 October 1960]. Note how this contrasts with the earlier-cited opinions of Bethe and Konopinski, to the effect that this was primarily a theoretical discovery, not an experimental one.

PAGE 151

1 H. Weyl, *Gruppentheorie und Quantenmechanik* (1931), 2nd edn., p. 234.

2 I. Tamm, *Z. Phys.* LXII (1930), 545; J. R. Oppenheimer, *Phys. Rev.* XXXV (1930), 939; P. Dirac, *Proc. Camb. Phil. Soc.* XXVI (1930), 361

3 J. R. Oppenheimer, *Phys. Rev.* XXXV (1930), 562.

PAGE 152

1 Millikan, *op. cit.* p. 320.

PAGE 154

1 $v = 1$ mile/sec.: Cf. *Mem. Manchr lit. phil. Soc.* (1951), 2nd ser., p. 107; *Phil. Mag* (1857), p. 211.

2 Maxwell, *Phil. Mag.* XIX, 28.

PAGE 155

1 Cf. Maxwell, 'Atom', in *Scientific Papers*, II, p. 480. Cf. *infra*, p. 194, n. 5.

2 Lodge, *Modern Views of Electricity*.

PAGE 156

1 Maxwell, *A Treatise On Electricity and Magnetism*, 'Electrolysis', 1873.

2 That is, in 1871, Weber, *Werke*, IV, 281.

PAGE 157

1 Stoney, 'On the physical units of Nature', *Phil. Mag.* XII (1881), 384.

2 Helmholtz, *Wissenschaftliche Abhandlungen*, III, 69.

3 Kelvin, 'Contact electricity and electrolysis', *Nature, Lond.*, LVI (1897), 84.

4 Stoney, *Sci. Trans. R. Dublin Soc.* 2nd ser., IV (1891), 563.

PAGE 158

1 Nernst, *Theoretische Chemie*, pp. 197, 456.

2 And still is, of course. Compare Chadwick, Blackett and Occhialini, 'Some experiments on the production of positive electrons', *Proc. Roy. Soc.* A, CXLIV (1934): 'The name "positron" for a particle of about electronic mass and positive charge was suggested by Anderson, and it seems to be coming into general use. The negative will still be called "as usual" an electron' (p. 235).

PAGE 159

1 Consider Schrödinger's characterization of the electron as constituting an electrical 'density smear' of interfering electromagnetic waves. *Ann. Phys.* LXXIX (1926), 361 and 489. Cf. also Dirac, writing in 1932(!), where he says: 'The [electromagnetic] field should appear in

the [relativistic quantum] theory as something more elementary and fundamental [than the particles].' *Proc. Roy. Soc.* A, cxxxvi, 454.

2 Reactions to the interpretation offered here have not been uniformly favourable. Thus, Oppenheimer writes: 'I, for one, am wholly unaware of the need for any explanation reaching back into history for a firmly held belief that there were only two particles in nature. We were talking about neutrinos and neutrons, and no one could look at the cosmic ray pictures without getting the impression that something was up that could be neither electron nor proton.' [Letter to the author, 12 August 1960.] Niels Bohr writes: '...at the time of the discovery...physicists were certainly prepared for surprises to a degree far beyond that prevailing in former centuries. Furthermore since quantum theory gives no clues as to which kinds of particles may exist in nature, the question of new particles was from the very beginning realized as completely open.' [Letter to the author, 28 October 1960.] And N. F. Mott writes: 'I would say looking back on it all that one's reluctance to accept Anderson's discovery was due to our lack of realization that the positive electron was created and unstable.' [Letter to the author, 26 August 1960.] Indeed, in light of Rutherford's anticipation of the existence of an elementary nuclear constituent, Bohr feels that 'it is hardly justified to ascribe to him a limited outlook influenced by the early theories of electricity'.

Against such an assembly one can barely imagine a defence. Yet the following remark of Professor Dirac is defence enough: 'I find this [interpretation] very good. It explains pretty closely my feelings at the time and my unwillingness to postulate a new positively charged particle.' [Letter to the author, 1 November 1960.]

With the issue left thus in equipoise, the reader may reach his own conclusions. I would add, however, that the postulation of the neutrino between 1923 and 1931, and the discovery of the neutron in 1932, seem not to count against the interpretation being tendered here. Neither particle is charged: our argument is primarily concerned with the possibility of there being a third charged particle, since it would have been commonly assumed that no fundamental particle could be neutral inasmuch as electrical neutrality was felt to result only from the balanced combinations of positive and negative particles. Besides, just citing the dates of the neutrino and the neutron will not by itself destroy my point, namely, that these particles were looked upon as being 'strange', and very difficult to absorb in any existing conceptual framework. Even so, Dr John Ziman rejoins: 'I just don't believe it was as strong as [your interpretation suggests]....All scientists are conservative. They hate to have to think about new elements....' [Letter to the author, 20 October 1960.] And Professor Bethe writes: '...true enough...but equally important was the desire for a minimum number of fundamental particles altogether...physicists were strongly opposed to any kind of inflation....' [Letter of 18 October, 1961.]

3 *Proc. Roy. Soc.* A, February 1933, pp. 699 ff.

4 Henceforth, the designations 'positron' and 'negatron' will designate positively and negatively charged electrons respectively. 'Positive electron' and 'negative electron' will designate the two fundamental units of electrical charge. This *may* avoid some of the terminological confusions of early twentieth-century electron theory.

5 Skobeltzyn, *C.R. Acad. Sci., Paris,* cxcv (1932), 315; Anderson, *Phys. Rev.* xli (1932), 405; Blackett and Occhialini, *Nature, Lond.,* cxxx (1932), 363.

PAGE 160

1 Skobeltzyn, *Z. Phys.* liv (1929), 686; *C.R. Acad. Sci., Paris,* cxciv (1932), 118; Auger and Skobeltzyn, *C.R. Acad. Sci., Paris,* clxxxix (1929), 55.

PAGE 161

1 Cf. Bethe, *Z. Phys.* lxxvi (1932), 293; also Carlson and Oppenheimer, *Phys. Rev.* xli (1932), 763.

2 Concerning terminology, Blackett writes: 'By a positive electron is meant a particle with unit positive electronic charge, and with a mass very much less than that of a proton' (*ibid.* p. 708). This does not accord with our decision above, note 4 (four notes above this), and the difference should be borne in mind.

PAGE 162

1 In his letter 'Cosmic-Ray Bursts', Anderson describes 'associated' tracks which are probably positron-negatron pairs. *Phys. Rev.* xliii (1933), 368. In *Phys. Rev.* xliv (1933), Anderson is 'dead on target' concerning pair-production in cosmic ray showers (p. 416). Cf. *infra* quotation preceding section B, p. 143. The final and still generally accepted theory of this process is set out in a paper by Heitler and Bethe, *Proc. Roy. Soc.* 1934.

PAGE 163

1 Cf. Millikan, *op. cit.* p. 335.

PAGE 164

1 Cf. Blackett and Occhialini, *Phys. Rev.* xliii, 1034; xliv, 411.

2 For example, the anomalous absorption of γ-radiation by heavy nuclei was now thought (by Blackett) to be connected with the formation of positive electrons and the re-emitted radiation with their disappearance. Cf. Meitner and Hupfield, *Naturwissenschaften,* xix

(1931), 775; Chao, *Phys. Rev.* xxxvi (1930), 1519; Gray and Tarrant, *Proc. Roy. Soc.* A, cxxxvi (1932), 662.

3 Cf. Chadwick, Blackett and Occhialini, *Nature, Lond.*, cxxxi, 473; Curie and Joliot, *C.R. Acad. Sci.*, *Paris*, cxcvi, 1105; Meitner and Phillipp, *Naturwissenschaften*, xxi, 286.

4 Cf. also the important paper by Oppenheimer and Plesset, in which it is shown '...in a particularly clear way, that the production of pairs is a typical quantum effect'. *Phys. Rev.* xliv (1933), 54: the Fermi–Uhlenbeck paper is in the same volume, pp. 501–11.

5 Cf. Chadwick, *Proc. Roy. Soc.* A, cxxxvi, 692.

6 Yukawa, *Proc. Phys.-math. Soc. Japan* (3), xvii (1935), 48, and esp. 57.

PAGE 165

1 First mentioned by Dirac, *Proc. Roy. Soc.* A, cxxxiii (1931), 62. Indeed, Anderson himself writes (in the original paper): '...the discovery of the positron should prove a stimulus to search for evidence of negative protons' (*ibid.* p. 494).

PAGE 166

1 In John Couch Adams's copy of the *Opticks* these last three words are crossed out, and in their stead has been inscribed 'to arise and be propagated'.

PAGE 171

1 Save after the technique of 'renormalization' is invoked in quantum field theory—which is one of the main objections to that technique.

2 $\oint F.ds = 0$, so that $F = -\nabla V$. Then H is the total energy of the generalized co-ordinate transformations, $rm = r(q_1 \dots q_n t)$—where q are generalized co-ordinates—do not depend explicitly on time.

PAGE 179

1 In mehreren Fallen ist bemerkt worden, dass die unstetige Änderung der Bewegungsrichtung von einer spontanen, manchmal sehr bedeutenden Änderung der Grösse der Geschwindigkeit begleitet ist. Diesbezügliche Betrachtungen, sowie die statistischen Angaben über die Häufigkeit der Einzelstreuprozesse werden später mitgeteilt.

PAGE 182

1 But if you finally agree with me, the publication of this letter (1956) will be purposeless.

INDEX

action at a distance, 155, 184, 186
ad hoc methods, 122, 147
Adams, 27, 36, 225
Alexandri Aphrodisiensis, 185
Alexandrow, 210
algorithm, 118ff., 213, 214
Alhazen, 8
Alice, 110
Almagest, 34
α particles, 107ff., 141, 164, 198
'α star', 107
amber, 153
American physics, 165
Ampère, 156
Anderson, 3, 24, 41, 56, 69, 70, 75, 92, 106, 122, 133–4, 135ff., 139ff., 146, 151, 152, 153, 159, 160, 163, 164, 165, 180, 182, 183, 198, 208, 216, 217, 218, 222, 223, 224, 225
Ango, 187
annihilation, 134, 151, 163, 164, 167, 185, 221
anomalous moment of the electron, 169
anomalous Zeeman effect, 198
anti-electrons, 75, 134, 151, 165
anti-matter, 92, 134, 165, 219
anti-neutron, 75, 165
anti-particle, 3, 75, 107, 151, 152, 164–5
anti-proton, 75, 152, 165, 225
Apian, 27
Apollonios, 26
Appendix I, 15, 166
Appendix II, 57, 167
Appendix III, 121, 171
Appendix IV, 137, 179
Arago, 191
Archimedes, 184
Aristarchus, 214
Aristotle, 26, 186, 193
'associated' particles, 142, 224
astronomy, 192
atomic explanation, 44ff., 189, 194, 195, 196, 201, 222
atomic particles, 42ff., 75, 154, 155, 195, 196, 198, 217, 218
atomic spectra, 168

atomism, 50ff., 189, 194, 195, 196, 205, 222
atoms of electricity, 154–6, 157, 196
Auger, 224
auxiliary hypotheses, 213
average properties, 69
axes, 120

Bacon, 17, 50, 200
'bare' electrons, 168–9
Bartlett, 64
Beeckman, 48
Benton, 77
Beringer, 197
Berkeley, 50, 52, 53, 57, 59, 91, 165
beryllium, 164
Berzelius, 51
Bessel, 214
β-ray spectrum, 48, 195, 199,
Bethe, 91, 98, 136, 138, 139, 161, 219, 221, 224
bi-prism experiment, 9
binding force, 51
Biot, 10, 71, 188–9
Birch, 51, 200
bivalent atoms, 155
black bodies, 5, 11, 31
'black box' theory, 34, 37, 39, 40, 214
Blackett, 1, 3, 24, 41, 56, 69, 70, 75, 92, 122, 133–4, 135ff., 153, 159ff., 164, 208, 222–3–4–5
Bogoliubov, 215
Bohm, 2, 31, 64, 76, 91, 93, 94ff., 102, 103, 105, 106, 111, 112, 117, 192, 193, 196, 206, 208, 209, 210, 211
Bohr, 17, 65, 71, 74, 76, 77, 79, 80, 86, 89, 93, 94ff., 101, 102, 103, 107ff., 122, 123, 139, 142, 146, 152, 159, 195, 201, 205, 207, 208, 209, 211, 213, 223
Bohr Interpretation, 94, 95, 111, 112
Boltzmann, 28, 51, 64, 79, 80, 208
Bopp, 93, 97, 102, 105, 209
Born, 74, 75, 77, 119, 123, 127, 128, 130, 132, 133, 171, 201, 207
boron counter, 197
Boscovich, 71, 187, 189
Bose, 198

INDEX

229